SATURN V
AMERICA'S ROCKET TO THE MOON

EUGEN REICHL

4880 Lower Valley Road • Atglen, PA 19310

Other Schiffer Books by the Author:

Project Apollo: The Early Years, 1960–1967 (978-0-7643-5174-7)
Project Mercury (978-0-7643-5069-6)
Project Gemini (978-0-7643-5070-2)

Originally published as
Saturn V: Der Mondrakete by Motorbuch Verlag,
Stuttgart, Germany © 2015.
www.paul-pietsch-verlage.de

Translated from the German by David Johnston

Library of Congress Control Number: 2017953838

Cover designed by Molly Shields

Type set in Avenir LT Std/Univers LT 47 CondensedLt

ISBN: 978-0-7643-5482-3
Printed in China

Published by Schiffer Publishing, Ltd.
4880 Lower Valley Road
Atglen, PA 19310
Phone: (610) 593-1777; Fax: (610) 593-2002
E-mail: Info@schifferbooks.com
Web: www.schifferbooks.com

For our complete selection of fine books on this and related subjects, please visit our website at www.schifferbooks.com. You may also write for a free catalog.

Schiffer Publishing's titles are available at special discounts for bulk purchases for sales promotions or premiums. Special editions, including personalized covers, corporate imprints, and excerpts, can be created in large quantities for special needs. For more information, contact the publisher.

We are always looking for people to write books on new and related subjects. If you have an idea for a book, please contact us at proposals@schifferbooks.com.

CONTENTS

From the Super Jupiter to the Juno V 4
From the Juno V to Saturn I 9
Danger to the Saturn 13
The Question of the Upper Stages 15
Nova Versus Saturn 17
The Configuration Is Set 19
The Mission Mode 21
Alabama, Louisiana, Mississippi 22
Launch Site Cape Canaveral 25
The Logistics 26
S-I and S-IB 30
H-1: Military Engine for the First Saturn 34
F-1: The Most Powerful Engine in the World.... 36
RL-10: From the Centaur to the S-IV 41
J-2: The Engine for the Moon 42
Upper Stages of the Saturn: S-IV and S-IVB 46
Base Block for the Saturn V: S-IC 52
The Problem Child: S-II 56
The Instrument Unit: The Brain of the Saturn ...61
First Flight and Teething Troubles 64
Epilogue for the Saturn 66

Specifications, Dimensions, Diagrams 68
Saturn I Block 2 (SA-5) First Stage 69
Saturn I Block 2 (SA-5) Second Stage S-II 70
Saturn IB (SA-201) First Stage S-IB 72
(SA-201) Second Stage S-IVB and IU 73
Saturn IB (SA-201) Payload Fairing 74
Saturn V First Stage S-IC 76
Saturn V Second Stage S-II 77
Saturn V Third Stage S-IVB and IU 78
Saturn V IU and Payload Fairing 79

The 32 Missions of the Apollo Program 80
Saturn I Block 1 – SA 1 80
Saturn I Block 1 – SA 2 82
Saturn I Block 1 – SA 3 84
Saturn I Block 1 – SA 4 86
Saturn I Block 2 – SA 5 88
Saturn I Block 2 – SA 6 90
Saturn I Block 2 – SA 7 92
Saturn I Block 2 – SA 9 94
Saturn I Block 2 – SA 8 96
Saturn I Block 2 – SA 10 98
Saturn IB – SA 201 100
Saturn IB – SA 203 102
Saturn IB – SA 202 104
SA 501 (Apollo 4) 106
SA 204 (Apollo 5) 108
SA 502 (Apollo 6) 110
SA 205 (Apollo 7) 112
SA 503 (Apollo 8) 114
SA 504 (Apollo 9) 116
SA 505 (Apollo 10) 118
SA 506 (Apollo 11) 120
SA 507 (Apollo 12) 122
SA 508 (Apollo 13) 124
SA 509 (Apollo 14) 126
SA 510 (Apollo 15) 128
SA 511 (Apollo 16) 130
SA 512 (Apollo 17) 132
SA 513 (Skylab 1) 134
SA 206 (Skylab 2) 136
SA 207 (Skylab 3) 138
SA 208 (Skylab 4) 140
SA 210 (Apollo–Soyuz Test Project) 142

FROM THE SUPER JUPITER TO THE JUNO V

After a very good beginning, things were not going well for Wernher von Braun. The Army Ballistic Missile Agency, or ABMA for short, had been founded on February 1, 1956. Its task: to develop ballistic missiles for the US Army. Maj. Gen. John Medaris was placed in charge of the new institution. After getting the Redstone short-range ballistic missile into service, the new authority's first big project was the Jupiter medium-range missile, development of which had begun three years earlier as a joint Army-Navy project.

From left: ABMA commander Maj. Gen. John Medaris, technical director Wernher von Braun, and the deputy commander of the ABMA, Brig. Gen. Holger Toftoy

The year before the formation of the ABMA, von Braun and Medaris had proposed to the Defense Department that a satellite carrier be developed from the Redstone and that it should put the first earth satellite into orbit before the end of 1956. However, the navy and air force intervened and the authorities subsequently prohibited the ABMA from continuing work on this project. The development contract for America's first satellite instead went to the Vanguard program of the US Naval Research Laboratories. Then in November 1956, the Jupiter program was also taken away from the ABMA and transferred to the air force. This led to fears that the army would soon have nothing more to do with large rockets.

Von Braun and his team were initially left with just the development of the Redstone. While it was designated the Jupiter-C, it had little to do with the Jupiter. Instead it was a version of the Redstone which was to be used to test warheads for the Jupiter medium-range missile. Another project on which von Braun and his team worked was designated Redstone-S. It was a successor to the Redstone, powered by solid fuel rockets, from which was later developed the Pershing battlefield missile. With this manageable portfolio, designed for short-range missiles, the navy and air force had their competitor exactly where they wanted it: limited to the development of tactical weapons for use by the army. This was not a situation von Braun could be satisfied with. It is only thanks to an irony of history that two years later von Braun put the first American satellite into orbit in a crash program after the failure of the Vanguard project group.

But the ABMA still had an ace up its sleeve, which secured its right to exist for the time being, for it developed concepts for a missile which was larger than anything the air force and navy actually needed at that time. This was Wernher von Braun's chance and he intended to use it.

The first studies for this large missile had begun in April 1957. The project's stimulus had come from the Defense Department, which was planning "space-supported installations" for communications, as well as space probes and weather satellites. This required rockets which were much more powerful than any in existence at that time. What was needed was a launch vehicle which could put payloads of between nine and eighteen tons into low earth orbit and was capable of accelerating space probes weighing between 2.7 and 5.4 tons to escape velocity.

The goal was a booster with a first stage thrust of no less than 1,500,000 pounds of thrust. That was ten times the output of the Jupiter, which had made its first flight on March 1, 1957. With this measuring stick in view, the new big launcher was designated Super Jupiter.

The satellite carriers then in preparation were, like the Thor, the Juno II based on the Jupiter, and the Atlas C, supposed to be capable of placing payloads with a maximum weight of 3,085 pounds into orbit. There were already plans, however, to increase their capacity to about 9,900 pounds by 1962, by using high-energy fuels and newly-designed upper stages. This was a significant increase, but nevertheless it meant that even these versions would be unable to meet the Department of Defense's requirements. It was therefore obvious that an entirely new class of boosters would have to be developed in a very short time, and at the lowest possible cost.

The ABMA's first study proposed the use of a single engine with a thrust of 1,000,000 pounds. This rocket engine was designated the F-1. At that time North American's Rocketdyne Division had been working on feasibility studies for such an engine for two years. In view of the tremendous development advance it required, however, it was unlikely that it could become operational in the next five years. In fact, nowhere was there a rocket engine even close to being in this thrust class. Given the requirement for a launch thrust of 1,500,000 pounds of thrust, however, even one of these hypothetical F-1 engines would not have been capable of launching the new super booster. And so the ABMA planning group abandoned the idea with the single engine, as much as it would have simplified the design, and instead proposed using a cluster of four smaller rocket engines.

The proposal was based on the knowledge that, in addition to the F-1, Rocketdyne was also working on an engine in the 360,000 to 380,000 pounds thrust class. Just like the F-1, this engine existed only on paper and was designated the E-1. Four of these, together with something nebulously called "off the shelf tankage," would form the first stage of the new super booster. What was meant by "off the shelf" was simply a fuel tank with the largest possible diameter from the stable of existing military rockets, which could be lengthened and adapted for the Super Jupiter without incurring high development costs.

By that time—the autumn of 1957—the matter had achieved a certain degree of urgency but had not been given the highest priority. It did, however, give Wernher von Braun and his group in Huntsville, Alabama, food for thought; but then two events unexpectedly gave the program its initial spark.

The launches of Sputnik 1, 2, and 3 at four-week intervals represented the first event. It was immediately clear that they had been launched by mighty booster rockets, much more powerful than anything the USA had at that time. The second event was the failed launch of Vanguard 1, which was a traumatic event for the American public.

Key ABMA personnel. From left to right: Ernst Stuhlinger (sitting), Brig. Gen. Holger Toftoy, deputy commander of the ABMA, Hermann Oberth (foreground), who worked for the agency between 1955–59, Wernher von Braun, its technical director, and Eberhard Rees, his deputy.

President Dwight D. Eisenhower was now forced to act quickly, and so he ordered the formation of a new, small and capable organization dedicated to improving the nation's technological base. This was the ARPA, the Advanced Research Projects Agency. From the beginning the agency had very few people, initially no more than about eighty (today there are 240) including the "office girls" as Roy Johnson, its first director, rather disrespectfully dubbed them. Despite its small staff of personnel, the president gave the small organization great power. It could make quick decisions and in particular loosen up the dollars with which paper studies could be turned into hardware.

While ARPA was still at work building up its organization, in the spring of 1958, the ABMA continued its studies for the Super Jupiter with the E-1 engine. By July, the ARPA had established itself and reviewed plans for the new big rocket. The huge 1,000,000-pounds-thrust engine sparked its greatest interest, however. It was immediately clear that this futuristic rocket engine was incapable of meeting the requirement for the soonest-possible creation of a booster in the 1,500,000 pounds thrust class. The F-1 engine was quite obviously years away from operational capability. Things looked no better for the E-1, four of which would be needed for each rocket. With the Soviet lead, quick action was the order of the day, so the ARPA directed that only engines that already existed be used for the 1,500,000-pounds-thrust booster.

Wernher von Braun's team in Huntsville already had extensive experience with the Jupiter rocket's power plant,

the S-3D made by Rocketdyne. This engine also powered the air force's Thor medium-range missile. It was thus already flight-tested, produced 150,000 pounds of thrust for launch, and was available in large numbers. So he soon developed the idea of a cluster of eight S-3D engines.

Although there was still no formal agreement between the two organizations, ARPA and the ABMA worked together closely from the beginning. This was obvious in the consensual choice of the name for the new booster: Juno V. This followed the designations assigned to earlier concept studies by the ABMA for large, multistage rockets, which had been named Juno III and Juno IV. By using readily available components, the ARPA estimated that about sixty-million dollars (equal to half a billion dollars today) and two years of research and development time could be saved compared to the Super Jupiter with E-1 engines.

After this landmark decision the ARPA authorized the ABMA to begin development work on the Juno V and released preliminary funding. The first document created bore the ARPA contract number 14-59 and was dated August 15, 1958. It is of great historical

Wernher von Braun received the American Medal of Merit from Pres. Dwight D. Eisenhower, in January 1959.

The S-I stages for SA 1 (right) and SA 2, here in final assembly in Production Building 4705 of the Marshall Space Flight Center.

importance, for today the paper is regarded as the official starting point of the Saturn project. In part, the document states: "Initiate a development program to provide a large space vehicle booster of approximately 1.5 million pounds thrust based on a cluster of available rocket engines. The immediate goal of this program is to demonstrate a full-scale captive dynamic firing by the end of calendar year 1959."

At that time, the engineers in Huntsville were already working on improvements to the S-3D engine. It was trimmed to a higher thrust, operation was simplified and system reliability was improved. On September 11, 1958, Rocketdyne signed a contract to modify the S-3D so that a group of eight engines could operate in the planned new rocket. The changes were extensive and expensive. Rocketdyne required

half of all the money initially envisaged for the entire Juno V development. The modified S-3D engine was now designated the H-1, to identify it as part of the big booster's development line.

With the remaining funds, ABMA went to Huntsville to see what test facilities were available and how they could best design the rocket's tank configuration. The best solution would have been to simply modify one of the existing test stands in Huntsville. The biggest one there was the one for the Jupiter. It could accommodate a stage with a length of 50 feet, a diameter of 8.85 feet and a thrust of 165,000 pounds. The question of what was necessary to enable the test stand to accommodate a power unit with a length of 78.75 feet, a diameter of 20 feet, and a thrust of 1,500,000 pounds led to perplexed shaking of heads.

And finally the funding also had to suffice to get the job done on a whole scrabble bag of other activities. A thrust structure had to be developed and built which could accommodate the power of the eight engines.

Also needed were inspection and testing facilities, tools, and machines, and all this in king-size scale but at low costs. An examination of military depots and other government dumps was made to see if they could be plundered to keep costs down.

As ARPA contract 14-59 only required a full-scale captive dynamic firing and no booster rockets in flight configuration, the booster, which soon began taking form in the workshops in Huntsville, looked like a patchwork from the leftover ramp of a military depot and the bargain basement of the Redstone Arsenal. The necessary tank volume in particular was already causing headaches. The development and production of monolithic tanks with a diameter of twenty feet required the development of completely new work techniques and production facilities and exceeded every reasonable time and cost framework. Something else would have to be considered therefore, and somehow existing components would also have to be used for the fuel tanks.

From its earlier production contracts the ABMA identified numerous Redstone tanks with a diameter of 5.8 feet which had been sent back to the manufacturer because of poor quality and production errors. These were now sitting there, awaiting further use. There were also a few half-completed units and several in storage as spare equipment. The search finally also came upon several 5.5-foot-diameter tanks from the Jupiter medium-range rocket.

Necessity made the developers inventive. If they had already clustered the engines, why not do

Patchwork rockets: here the Jupiter tank, which formed the central unit of the S-I, is being attached to the hub assembly. On the far right will be the upper end of the rocket. The arrangement of many small nitrogen tanks can be seen well. On the right one of the Redstone tanks waits for installation. The photo was taken on January 25, 1960, in Building 4705 of the Marshall Space Flight Center.

the same with the fuel tanks? And so the planned super-modern booster turned into a curious compromise, in which eight Redstone tanks were arranged around a single Jupiter tank like the cylinder of a revolver. It no longer looked exactly like a streamlined, futuristic vehicle for the future conquest of space, but it also was not envisaged for that. This hodgepodge of spare parts and leftovers had been put together solely to find out if it was possible for the eight engines of a behemoth rocket to run simultaneously.

While work in Huntsville made good progress, representatives of the ARPA kept a watchful eye on the project's progress and made frequent visits to the Redstone Arsenal. While what they found there looked curious, it pleased them more and more. It pleased them so much that the proposal was soon made to derive a flight-capable version from this pure test stand machine. On September 23, 1958, the ARPA and the Army Ordnance Missile Command, the ABMA's umbrella organization, agreed to also carry out a flight test of the vehicle after completion of test stand experiments. As well, von Braun's group was to produce three more boosters, of which the last two were supposed to be capable of placing a limited payload into earth orbit.

FROM THE JUNO V TO SATURN I

On July 28, 1958, Pres. Eisenhower signed the National Aeronautics and Space Act, the founding document of the National Aeronautics and Space Administration, much better known by the abbreviation NASA. The new authority stood on the foundation of a predecessor organization that was already forty-three years old. That was the National Advisory Committee for Aeronautics, or NACA. On this basis, on the day it began operating, October 1, 1968, the new organization had almost 8,000 employees and four research centers.

None of these four centers had so far been involved in the development of large rockets, however. Only at the Jet Propulsion Laboratories in California was there a certain expertise in low-thrust rocket engines. The first administrator of NASA, Keith Glennan, therefore immediately wanted the ABMA team under von Braun in NASA. Unfortunately, the military's persistence would not allow this, and so the organization initially remained in the army, but not for long.

As one of many measures to counter the Soviet threat, in January 1958, the NACA created a special committee for spaceflight technology. The working group for space vehicles was headed by Wernher von Braun. He was at the same time also ABMA's representative in this group. His appointment, following his success with Explorer 1 just a few days earlier, was an obvious choice. This committee tabled its final report on October 28, 1958, or four weeks after NASA was created. Von Braun saw where the journey was leading, and he wanted to be part of this journey. It was completely clear to him that the army would not be in a position to realize his plans. Only the new NASA could do that. The new authority lacked expertise in large rockets and von Braun filled this gap perfectly.

Von Braun's report began with harsh criticism of the present status, where a number of competing, mainly military organizations were working in parallel with different concepts and against each other on the highest priorities of the national space program. He pointed out that the Soviet Union had definitely taken over the leading role in spaceflight and the militarization of space. And he showed a way to address this situation. Then von Braun described five generations of booster rockets.

First there was the Vanguard, which as a super-light booster was in a class of its own. The next stage consisted of the Jupiter C, the Juno II, which was derived from the Jupiter, and the Thor-Able, a combination from the air force's Thor medium-range missile and the Able upper stage from the Vanguard program. The boosters of the first two generations had one thing in common: both been hastily developed in response to the Sputnik launches by the Russians. They were only capable of carrying very small payloads and were extremely unreliable.

Generation three consisted of derivatives from the intercontinental ballistic missile program, namely the Atlas and Titan. Generation four was what von Braun called the clustered booster: boosters which were built according to the cluster principle and developed up to 1,500,000 pounds of thrust. It was precisely the project on which he was then working.

And finally he outlined the most massive rocket with a completely new generation of super-powerful single-chamber engines, two to four of which would be combined into one booster to produce a total thrust of up to 5,600,000 pounds.

But for what purpose did he intend to use these rockets? The working group simply speculated. A four-man space station could be built with the help of Atlas and Titan by 1961. Moon landings could take place by 1966, with the help of clustered rockets, just like a fifty-man space station by 1967. Finally, with the super booster, they wanted to support longer moon missions from 1972, a moon base from 1974, and manned interplanetary flight from 1977. But the primary target, according to the report, was first the development of clustered boosters and the start of the time-consuming development of the F-1 engine, the key element of the super booster.

On January 27, 1959, NASA delivered a report to Pres. Eisenhower in which a national booster

From the beginning, the Saturn program interested political leaders in the USA. Here Wernher von Braun, Pres. John F. Kennedy, and Lyndon B. Johnson in front of Saturn SA 1 at the Marshall Space Flight Center.

program was introduced. The author of this program was NASA engineer Milton Rosen. Rosen had been program leader for the Viking high-altitude research rocket and subsequently managed its development into the Vanguard satellite booster. His report included most of what Wernher von Braun had already brought up in the special committee for spaceflight matters but only three classes of new booster rockets.

The first was to be based on the Atlas and produced in two versions. One version, the one and a half stage standard model, would be capable of placing payloads of up to 1.4 tons into a low earth orbit. The other version would have a more modern upper stage powered by hydrogen and oxygen. NASA was already carrying out studies for this project in which the new rocket was called Centaur. With this new upper stage, they would be capable of placing into orbit payloads four times heavier than with the standard Atlas. In general, Rosen saw the mastery of liquid hydrogen technology as vital to the success of advanced spaceflight programs. Otherwise, he

continued, the payload capacities of future boosters would remain disappointingly low.

The second class was to consist of the Juno V with its eight clustered first stage engines. This booster was to be produced in two versions with different upper stages. Both were designed to be three-stage rockets. On the Juno V-A the second and third stages would simply be from the Titan I intercontinental missile, which was a two-stage booster. The second stage of the Juno V-B would also be a Titan I first stage, but the Centaur would be used as the third propulsion unit. The Juno V-A was to be used for flights to earth orbit, the Juno V-B for missions to the moon and the planets.

Finally, the third class was based on a completely new super rocket of extraordinary size and payload

The first Saturn I Block 1 in flight configuration. For comparative purposes the technicians of the Marshall Space Flight Center have placed a Mercury Redstone next to the Saturn and in the background a Juno II satellite booster. The photo was taken in early 1961.

capacity. In Rosen's report it was called Nova. Its propulsion system would be based on the 1,500,000-pounds-thrust engine. Four of these engines in the first stage would create 5,600,000 pounds of thrust at launch. The second stage was to use one of these engines and each of the third and fourth stages' hydrogen-powered Centaur engines, four in the third stage and one in the fourth stage. In Rosen's concept these rockets would have had a diameter of 32.8 feet and stand 260 feet tall. Rosen pointed out that despite its immense size, this booster was the minimum requirement for landing a man on the moon and bringing him back again.

It was also at about this time that the new big booster had its second name change. At first it seeped quite gradually into the project. Neither the name Super Jupiter nor Juno V had found much

favor with Wernher von Braun's team in Huntsville. From the beginning they wanted to give the new booster rocket a name that would identify it as something new and not simply as a derivative of the medium-range Jupiter. For the engineers of the Redstone Arsenal a catchy new name offered itself. After all, they had named the Jupiter after a planet in the solar system and so it was only logical to give the new rocket the name of the next planet, Saturn. The ABMA also had nothing against it. There was a tradition there of naming rockets after ancient gods, like Thor, Atlas, Titan, or even Jupiter. Saturn was the Roman god of agriculture and of the mythical golden age and fit this concept perfectly. So the name was officially changed in February 1959, and the Department of Defense announced that the Juno V program would henceforth operate under the name Saturn.

In a preparatory technology program for production of the Saturn I, here in May 1959, a new method is tested for circumferential welds around the entire tank.

DANGER TO THE SATURN

This made it all the more surprising when, in June 1959, the Department of Defense completely unexpectedly began its attacks on the Saturn program. Herbert York, director of the Defense Research and Engineering Department, announced that he was thinking of stopping the program. In a memo to ARPA director Roy Johnson, he had previously sharply rejected his request for additional funding: "With regard to the Saturn," he wrote, "it is my opinion that there are significantly more important projects which should be served from the limited budget." He was thus expressing an opinion that was widely held in the Eisenhower administration. The influential military was convinced that the goals of the Department of Defense could be achieved significantly better with special versions of existing intercontinental rockets. The Saturn program was increasingly seen as the ARBA's expensive hobby horse. York informed Johnson: "I have decided to stop the Saturn project, because it has no military justification." The news landed in the ABMA like a bomb. The first H-1 engine had arrived in Huntsville just a few weeks earlier for integration into the first Saturn I test stand model.

But it was not just the ABMA that was touched by the thunder. The news also caused dismay in NASA, for its future planning was closely tied to the Saturn. The management of the space authority immediately began taking measures at the last minute to prevent the program from being stopped. In cooperation with several advocates of the Saturn in the DOD (and there were some there) and with additional support from the ARPA it drafted a memorandum. In it, it acknowledged quite clearly that the army had no need for such large rockets. In fact NASA also had no requirement for them at that time. The Saturn was a promise for the future, not a necessity of the present, but it would have to be saved in the present of 1959.

Between September 16–18, 1958, decisive discussions were held about the future of the Saturn. NASA's leading negotiator was Deputy Administrator Hugh Dryden. The Department of Defense was represented by Herbert York. After hours of discussions, NASA's arguments tipped the scales and York relented. But the victory was not unconditional, for York insisted that the ABMA and the Saturn project must be transferred to NASA. And such a transfer only made sense if the space authority itself funded the Saturn program.

The near loss of the Saturn booster was a sobering experience for NASA. Quite obviously it needed control over its own booster rockets to be freed of such incidents in the future. The takeover by NASA did not proceed immediately, however. The reason for this hesitation was clear: NASA had no specific task for the Saturn to justify all of the financial and administrative responsibilities associated with it. To accomplish this, NASA created the Goett Committee, named after its leader Harry Goett. This committee was to work out the future of the USA's manned spaceflight program after the Mercury project. Within its study, circumlunar missions and landing on the moon were identified for the first time in NASA's long-range planning. Now the space authority had its target, still vague and in the distant future, but now the responsibility for the Saturn no longer made any sense anywhere else than in NASA itself.

The transfer of the ABMA including the Saturn project and Wernher von Braun's team took a period of almost six months. Authorization was given by Pres. Eisenhower on November 2, 1959. The transfer document stated that… "no clearly identifiable military requirement for a super booster exists, however a clear requirement for it exists within the civilian space program. Consequently it is agreed that NASA be entrusted with responsibility for the super booster program."

The last administrative step was a letter from Eisenhower to NASA Administrator Keith Glennan in January. In it he wrote: "You are hereby directed to make a study, to be completed at the earliest date practicable, of the possible need for additional funds for the balance of Fiscal Year (FY) 1960 and

for FY 1961, to accelerate the super booster program for which your agency recently was given technical and management responsibility." On March 15, the development complex within the Redstone Arsenal in Huntsville was incorporated into NASA under the name George Marshall Space Flight Center. This included the associated facilities covering an area of almost 1.9 square miles. On March 16, NASA also added the ABMA to its ranks. On July 1, 1960, all of the personnel of the Redstone Arsenal were formally transferred to the space authority. Thus 4,670 former employees of the US Army became employees of NASA overnight. Leadership of the new organization was given to Wernher von Braun.

The Special Committee on Space Technology was composed of top-class individuals. It included names such as Randolph Lovelace (fifth from left), Julian Allen, Robert Gilruth (at the twelve and one o'clock positions at the head of the table), Hugh Dryden (fourth from right), and Abe Silverstein (next to Wernher von Braun). An irony of history: Hendrik Wade Bode (fourth from left), who during the Second World War, developed radar-controlled artillery for use against V-1 rockets, sitting at the same table as Wernher von Braun, who at Peenemünde developed the V-2, which was used in terror attacks on London.

THE QUESTION OF UPPER STAGES

While politicians and leading managers were still thinking about the future organizational arrangement of the ABMA, the engineers were already busy, thinking about what the upper stages of the Saturn might look like. Only they would turn the booster, which was just being put together, into a powerful launch vehicle. The main lines of this concept bore the names Saturn A and Saturn B and were conceived in the early months of 1959. Their only common feature was the first stage with a diameter of 21.3 feet, on which work had already begun. There was complete bewilderment above that, and countless designs were created for upper stages, some as tall as 250 feet.

Initially the use of newly-developed hardware was not considered, following the guidelines of the ARPA and the ABMA for extreme thrift. In the late spring of 1959, after many iterations, a three-stage variant seemed to prevail, in which a second stage from the Titan program and the future Centaur upper stage from the Atlas program played a role. This solution was not entirely satisfactory, however.

There was a change in July. The Department of Defense directed a question at the air force and the ARPA, asking if they could not design a completely new upper stage for the Saturn. What it had in mind was a propulsion unit which could also be used for the air force's Dyna Soar project. At that point the ARPA halted the studies using versions of the Titan upper stage and decided to wait.

The ABMA was relieved. The Atlas and Titan upper stages might be suitable for their role as ballistic nuclear weapons carriers, but as upper stages for such a powerful booster as the S-I they were seriously underpowered. Even their dimensions did not fit well with the Saturn, for their diameters were limited to 9.85 feet for military reasons. It was a little like buying a five-ton truck and then using it to transport a wheelbarrow load of sand.

In any case, the theme of using existing hardware was off the table, especially as the air force decided that the Titan IIIC would be used to carry the Dyna Soar and not the Saturn.

Work on upper stage definition started again in autumn 1959. This time they began with a blank sheet of paper. There were only two remaining limiting factors. One was the maximum possible load capacity of the booster stage, the other NASA's future requirements. And it was at that time that Abe Silverstein entered the game.

Silverstein was head of development of the NASA spaceflight program and at that time headed the Saturn Vehicle Evaluation Committee, which went down in spaceflight history as the Silverstein Committee. In that period of uncertainty as to how

This contemporary drawing shows the first two stages of the so-called "building block concept."

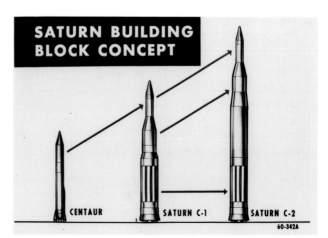

SATURN BUILDING BLOCK CONCEPT

CENTAUR   SATURN C-1   SATURN C-2

60-342A

to proceed with the Saturn upper stage, he seized the initiative and advocated the use of liquid hydrogen as fuel and liquid oxygen as oxidizer. In the 1950s, this combination was considered technically difficult to control, but compared to engines propelled by RP-1 this combination promised 40 percent greater thrust performance.

Through the entire 1950s, at the Lewis Research Center there was a technology program led by Abe Silverstein which occupied itself with the possibilities of liquid hydrogen as a rocket propellant. This research had culminated in a successful test of a combustion chamber prototype with a thrust performance of 20,000 pounds of thrust in the late 1950s. Despite this experimental model, hydrogen-oxygen propulsion was still largely unexplored territory. Wernher von Braun was anything but enthusiastic about Silverstein's advance and initially was unconvinced.

Silverstein had an easier time with other decision makers, which was probably due to their ignorance of the technical hurdles that still lay before them. Roy Johnson of the ARPA called the hydrogen-oxygen stage the Miracle Stage and his institution was determined to support the project.

The first practical use of a hydrogen-oxygen drive was the planned high-energy upper stage for the Atlas with the designation Centaur. General Dynamics had received a development contract for this system from the air force on November 14, 1958. Pratt & Whitney was to deliver the engine and it was now put forward for use as the Saturn's upper stage.

At the time of the Silverstein Committee all of the Saturn concepts were still three- to four-stage systems. Under discussion were the A-1, A-2, and B-1 variants, which in part required the development of new conventional stages. At this point an obvious idea came into play: why should liquid hydrogen be limited to the upper stage as a propellant? If it was accepted, then there was no reasonable reason why it should not also be used for the stages in between. The requirements were rather less, because here aspects such as a possible re-ignition in orbit were not a factor.

These ideas now formed the basis of the C version of the Saturn family. The program lines A and B, which had been under investigation, were now dropped for good. Defined in the new line were the C-1, C-2, and C-3 versions. All of the stages, apart from the base unit, were now powered by liquid hydrogen. And the base stage for all three was the S-I, which was already in development.

The numbering did not simply indicate the different variants, rather it reflected the evolutionary nature of the development project. Silverstein called these development steps building blocks. Each subsequent stage of evolution was based on its predecessor. The three-stage C-1 was the base version for all subsequent versions. The first stage was the S-I booster, the second stage was a new design but used four Centaur engines, and the third stage was an "original Centaur" stage taken unchanged from the Atlas missile program with two of the new Pratt & Whitney engines.

As soon as the C-1 was flying, development of the C-2 would start. A new stage, called the S-III, was to be developed for this evolutionary step. It was to receive a single newly-developed hydrogen-oxygen engine with ten times the thrust of the Centaur engine.

As soon as the Saturn C-2 with the new S-III stage began flying, a new S-II stage with four of

This contemporary drawing shows the C-3 variant of the Saturn, still being planned at that time, and two versions of the Nova, which was considered indispensible for the direct flight method of a moon landing.

these high-performance engines was to enter service. Then all of these innovations would be brought together in the C-3: the base stage with its eight H-1 engines, the S-II second stage with its four new rocket engines, then the S-III with the single new engine and finally a Centaur variant with two engines. Last but not least, Silverstein proposed that the development of new, large hydrogen-oxygen engines for the S-II and S-III stages begin immediately.

NOVA VERSUS SATURN

Today, President John F. Kennedy's famous speech to Congress on May 25, 1961, is seen as the starting point of the American manned moon program. This is correct insofar as it gave NASA the mandate and the resources to undertake an accelerated development program with the highest priority, to counter the obviously growing Soviet lead in space. That the moon was NASA's goal from the beginning,

This photo was taken at Cape Canaveral in 1959. To Wernher von Braun's left is Air Force Gen. Donald Ostrander, and to his right Eberhard Rees, von Braun's deputy, and Brig. Gen. John Barclay of the ABMA.

however, was never in question after the Goett Committee's report at the latest.

In January, NASA's launch vehicle director, Maj. Gen. Donald Ostrander, announced at a planning conference for industry: "The primary goal ... is that of a manned landing on the moon and the (safe) return (of the crew) to earth." At a congressional hearing the same month, the NASA managers informed the congressmen of a just-completed ten-year plan. This plan envisaged carrying out the first circumlunar mission by about 1970, and not too long after that landings on the earth's satellite. Ostrander named three rocket types with which they intended to achieve this goal: the Saturn types C-1 and C-2 and a new rocket called Nova. The latter would be used for the moon landings. The Saturn was not powerful enough for this purpose.

No one thought of questioning the necessity of an aggressive plan and an ambitious target. Because of constant Soviet pressure, the space program had already undergone considerable acceleration under Eisenhower. On January 18, 1960, the Saturn program received the DX rating, enjoyed only by programs with the highest national priority. This simultaneously authorized large-scale overtime, the associated higher wages for personnel and priority with respect to resources compared to other government programs. The Saturn program had gone far in a short time, from near cancellation to absolute national priority in just eighteen months.

Under the new president it quickly gained more speed. Reports of Russian successes came one after another, and the feeling of Americans that they were being beaten in the struggle against the class enemy, grew steadily. On April 12, 1961 just a few weeks after James Webb and Hugh Dryden had been sworn in as the new leaders of NASA, the Soviets put the first man into earth orbit, Yuri Gagarin. On May 25, Kennedy made his famous speech to congress and promised to increase NASA's budget by 500 million dollars for the year 1962. Allowing for inflation, this is equal to four-billion in present day dollars.

With the exception of the C-1, in early 1960, all other versions of the Saturn family were still under discussion. This was primarily due to the as yet uncertain mission mode for the Apollo moon landing ship. In those days the direct flight to the moon was favored. In terms of mission scenario, this method was the simplest. No multiple launches or rendezvous maneuvers were required. All other mission modes were based on one or more of the many rendezvous variations, and with one exception all required two or more launches per moon landing.

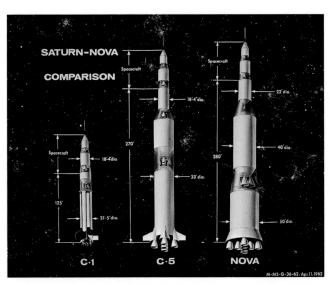

This drawing was made a year after the picture on page 16. Development of both the Saturn concept and the Nova has progressed.

The direct flight method, however, required a super-heavy launch vehicle. At the beginning of June 1961, NASA calculated that in the case of a direct landing, a mass of ninety tons would be required to achieve escape velocity. For a launch vehicle with this power, it was calculated that the first stage would require no less than eight of the new F-1 engines. It would be a completely new program, and not least for this reason the rocket was dubbed Nova. It had a launch thrust of 12,365,000 pounds of thrust with a launch weight of 4,800 tons.

As well, in addition to the J-2, which alone was ten times as powerful as the RL-10 upper stage engine, another hydrogen-oxygen engine, again six times as powerful, would have to be developed for the second stage. The Nova would have to be equipped with four of these engines, dubbed the M-1, in its second stage. Its total thrust would then be 4,945,000 pounds. A single M-1 engine would power the third stage and then a fourth stage would be required, equipped with a single J-2 engine.

After Kennedy's speech in 1961, the question of the mission mode soon took care of itself. Design of the Nova would have taken years longer than development of the most powerful Saturn variants. Kennedy's goal of a moon landing before the year 1970 would not be achieved. There were also objections, like those of Wernher von Braun, who warned against building the Saturn and the Nova simultaneously. This, he claimed, would far exceed NASA's resources. The Nova finally disappeared into oblivion after 1962.

THE CONFIGURATION IS SET

At the time of Pres. Kennedy's historic speech, the S-I stage of the Saturn C-1 was about to make its first test flight. The plan envisaged adding the S-IV second stage after the fifth mission. In January 1961, the C-1, previously planned as a three-stage rocket, had been made into a two-stage one. The S-V upper stage—no more than a Centaur adapted for the Saturn—was thus eliminated. Development of the S-V was resumed in February however, this time to use it as the third stage of the planned C-2 version.

Basically everyone had a say, and the configuration changed from week to week. Even the experts of the day were soon unable to keep up with the multitude of new variants. The constant changes were due to the fact that throughout all of 1961, the Apollo mission mode remained uncertain. The only thing that was halfway certain, was that it would probably be one of the rendezvous methods. Also certain: to cover all rendezvous methods, but also to carry out Apollo test flights in earth orbit, in the circumlunar area and in lunar orbit, they would have to prepare a mix of different powerful versions. For this reason consensus was soon reached on a more powerful version of the Saturn I, which was designated the Saturn C-1B. The base stage was the same as that of the Saturn C-1, but the H-1 engines now had somewhat greater thrust, and the J-2 engine, which had just begun trials, was to be tested in the second stage.

In February 1961, the C-2 was conceived as a three-stage version of the C-1B for missions requiring earth escape velocity. It would have been given a chance if NASA had decided on the Lunar Surface Rendezvous. This version quickly showed itself to be underpowered, however, for the weight requirements coming from the Apollo program rose steadily and sharply.

And so in June 1961, the C-2 was cancelled in favor of the C-3. Silverstein had already recommended this version in his report. It was to have three of the mighty F-1 engines in the base

While the question of future upper stage configurations was still being discussed, the first S-I flight stage for the SA 1 mission was completed at the Marshall Space Flight Center.

stage, four J-2 engines in the second stage, and four Centaur engines in the third propulsion unit. The C-3's concept life was little longer than the C-2's however, for it too proved to be too small.

The C-4 enjoyed a brief career in the summer of 1961. It had four F-1 engines in the first stage, four J-2s in the second propulsion unit, and a one in the third stage. It was to be used both for circumlunar missions and lunar orbital flights. Two C-4s could also accomplish a moon landing using the earth rendezvous method.

The C-4 had a serious chance of being realized, but gradually the view took hold that the largest possible launch vehicle should also be developed for the rendezvous mode, in order to reduce the otherwise necessary multiple launches to a few—ideally no more than two. Another consideration was that the Apollo program was still in the beginning stages and it was therefore impossible to say what other demands might be placed on the launch vehicle during development. If they began too small, it might have fatal results in later years. They simply needed flexibility. And so the trend gradually shifted from the C-4 to the C-5. This launch vehicle was to be equipped with five F-1 engines in its first stage, five J-2 engines in its second stage and a single J-2 in the third propulsion unit. The C-5 would cost more to develop and build, but with it any desired rendezvous method could be developed.

THE MISSION MODE

Until Kennedy's speech, the question of the mission mode had scarcely been discussed. The answer seemed both clear and simple: fly directly to the moon and land there. The disadvantage of this so-called direct flight method was that it was by far the most challenging in terms of energy technology. Wernher von Braun personally supported earth orbit rendezvous. It required the spacecraft to be assembled from several components placed in earth orbit by smaller launch vehicles. The disadvantage of this method was its complexity. Two or more individual modules had to be placed in orbit in a short time, where they first had to locate each other and then dock. At the beginning of the 1960s, that appeared difficult and impractical.

The earth orbit rendezvous was, however, far from being the only possible rendezvous method. NASA looked into many different procedures. Lunar Surface Rendezvous had a relatively strong group of adherents. It shifted the finding of the target vehicle to the lunar surface and avoided the unpopular docking maneuver. The basic idea was that a number of unmanned vehicles should be sent to a selected location on the moon. One of these vehicles would

have been the return vehicle for the astronauts. The others were tankers and supply vehicles. Using automatic techniques, these supply units would have fueled the return stage. As soon as it was operational, an individual astronaut would set out from the earth, land on the moon next to the fueled return vehicle, climb aboard it, and then return to the earth.

Other methods were examined, such as the Lunar Transfer Rendezvous. It envisaged the transfer of fuel during the flight to the moon. Such a measure had a mass advantage compared to the conservative orbit rendezvous methods, but it was soon abandoned as too unreliable and dangerous.

While all of these more or less exotic proposals were being discussed, a new plan came up. At first glance it looked even crazier than the other methods and it initially caused NASA planners to shake their heads. The new method was Lunar Orbit Rendezvous. Here the rendezvous maneuver would take place in lunar orbit and not in earth orbit. From the weight aspect this method was by far the best variant, however the idea of complex search and docking maneuvers in lunar orbit, 250,000 miles from Mission Control, caused shivers to run down the backs of the planners.

The man who forcefully supported this idea was John Houbolt. He proposed that part of the spacecraft, which would return to earth, be parked in lunar orbit before the start of the actual landing operation. This unit, he argued, would not be needed for the landing. Why then take it to the lunar surface and then transport it back up again? Instead, a specialized

John Houbolt explains Lunar Orbit Rendezvous.

vehicle, as small as possible, would carry out the landing and subsequent ascent. Fuel consumption would drop by an entire magnitude compared to the direct flight method.

For almost all of 1961, Houbolt tried to sell his idea. He saw himself, as he later said, as "a voice crying in the wilderness." Resistance within NASA was no surprise, for almost everyone, Wernher von Braun included, viewed the method as highly risky. The rendezvous, locating a target in space, was seen as a difficult task even in earth orbit, to say nothing of in orbit around the moon. The page did not turn until, in June 1962, the supporters of lunar rendezvous succeeded in drawing Wernher von Braun to their side. He still thought it dangerous, but in the end allowed himself to be convinced that

given President Kennedy's approaching target date it was the best of all the procedures. In particular the method had one undeniable advantage: lunar rendezvous allowed the moon landing to be carried out with one Saturn C-5.

The fog gradually cleared, and with it the confusing nomenclature. The Apollo program would be developed with the Saturn Types C-1, C-1B, and C-5. NASA made the decision in favor of lunar rendezvous on July 10, 1962. In February 1963, NASA headquarters announced the new designation for its launch vehicles. No further variants were being examined, and so they could end the C line as boundary to the A and B variants. The C-1 would simply be the Saturn I. The Saturn C-1B became the Saturn IB, and the Saturn C-V became the Saturn V.

ALABAMA, LOUISIANA, MISSISSIPPI

Design and construction of the Saturn I first stage began under the direction of Wernher von Braun in the old facilities of the US Army's Redstone Arsenal. The Juno, a booster derived from the medium-range Jupiter, was more or less the largest that the infrastructure there could accommodate, but now, at the very start of S-I development, a series of technology problems had to be solved. For this they would have to creatively cope with the existing facilities and simultaneously begin the necessary expansions.

There was, for example, the fundamental question of the ignition procedure for the engines in the S-1 stage. In these early days of spaceflight it was an accomplishment if a single engine was made to run reliably. The Saturn I, however, had eight, which had to be ignited in a precisely-determined choreography and then remain in stable operation. To be able to conduct experiments in Huntsville, an existing Juno test stand was modified. Then the facility had two test positions, one for the

Juno and one for experiments with the Saturn I. The modification was extensive and took almost a year. The test stand needed a new concrete foundation, a new flame shaft, new instrumentation and a huge crane capable of lifting 120 tons. The first engine runs with a test booster, designated SA-T, began in February 1960. Initially they ran just two engines for eight seconds. On April 29, 1960, all eight engines were run together for the first time, but only for a few seconds. It was mid-1960 before the entire cluster achieved a burn time of two minutes.

Modifications such as this were only a makeshift solution, however. They would never be adequate for tests with the S-IC and S-II stages of the Saturn V. And so huge new facilities were built, not far from the Redstone Arsenal. Coming from the north, from the gentle hill landscape of rural Alabama, one could see from far away the structure which extended for miles along the river. In the foreground were the old facilities of the ABMA with their laboratories, research centers, and workshops, extending several

The subsequent static firing test of the S-I stage for the SA 1 mission at the converted test stand for Juno II rockets at Huntsville. In position on the right is a Juno II. The photo was taken at the beginning of 1961.

blocks. Then came the new administration area and the new production buildings. And below on Indian Creek, a tributary of the Tennessee River, the silhouettes of gigantic research facilities dotted the countryside. Located there were the engine test stands and the facilities for turbopump experiments. Two structures in particular stood out. One was the big test stand for the first stage of the Saturn S-IC. It was completed in 1964, and stood 403 feet tall. Even larger was the Dynamic Test Stand, which was 423 feet tall. It was designed to accommodate a complete Saturn V with all stages, including the spacecraft.

The residents had been used to testing of small rockets (and the Juno II satellite booster was considered "small" in this terminology) since the mid-1950s. The acoustic footprint of a Saturn test was in an entirely different league, however. The result was large numbers of broken windows and flaking wall plaster in Huntsville and the surrounding area, followed by numerous compensation claims by the residents. NASA paid promptly and efficiently, which resulted in claims coming in even when there were no tests carried out. It was soon realized that the volume depended in large part on the weather. Tests often had to be delayed in cloudy or rainy weather, when the noise was especially well conveyed.

This of course could not become a permanent condition. The facilities in Huntsville were near a populated area and with the rising production numbers and the development of even larger stages the noise soon exceeded acceptable levels. A plan was therefore soon developed to carry out the majority of the trials, especially the production acceptance tests, at another site, if practicable near the production sites. And it found its home in Michoud, an eastern suburb of New Orleans.

Michoud was a special subject. Huge manufacturing halls were needed to build the first stage of the Saturn I and its successors, still on the drawing board. For time reasons, one that already existed was preferred. Proximity to the national waterways was absolutely vital, in order to be able to move the huge devices. The tremendous size of the launch vehicle and the necessity for accessibility between the manufacturing plants, test facilities, and launch site, brought an additional limiting factor into play: they did not just need some sort of waterway, they needed canals and sea routes which were guaranteed to be ice-free all year long. The logical choice for this combination of factors was the southern part of the USA, in particular the Gulf Coast. It had a mild, snow- and ice-free climate and a well-developed system of canals with access to sea routes.

As far as production sites were concerned, there were two acceptable sites. Both were mothballed large facilities from the Second World War, which were practically in possession of the government. One was near St. Louis, the other in Michoud, a small town fifteen miles east of the center of New Orleans. St. Louis was eliminated because the Mississippi regularly froze there in winter. So the choice almost automatically fell on Michoud. Located there was a truly mammoth building with a covered area of 1,830,000 square feet. It was part of a complex 1.35 square miles in size situated directly on the water. During the Second World War, the facility had first been a shipyard, then a factory for transport aircraft, and during the Korean War a factory for tank engines. The complex was ideally suited and Wernher von Braun selected it on September 7, 1961. NASA and the Chrysler Company would build the first stages for the Saturn I and the Saturn IB, and Boeing the first stage of the Saturn V.

The selection of Michoud was also influenced by the fact that, in addition to assembly buildings, test facilities of previously unknown dimensions—close by and of course accessible to ships—were also needed. NASA wanted to be able to carry out acceptance trials of the rockets produced in the factory as close as possible to the production sites. Safety and noise reduction regulations demanded a huge area, which could not just accommodate the test stands, but also serve as an acoustic buffer zone. Such an area was located in the border area between Mississippi and Louisiana, about forty miles from the factory in Michoud. Specialists had calculated that an area of 155 square miles would be required for this buffer zone.

NASA initially purchased twenty-one square miles of land and secured the purchase rights to an additional 200. To ensure that the area was completely unpopulated, about 100 families had to be relocated, including the entire village of Gainsville, Mississippi. Nevertheless there were no significant protests. NASA paid high prices and most residents were happy about the numerous employment opportunities that were created in this remote area. The central large facility of this new Mississippi Test Facility (MTF) complex was a huge test stand for the S-IC stages with two test positions and two separate test stands for the S-II stages.

Unloading the S-IC-5, earmarked for the Apollo 10 mission. Here it has been delivered by barge to the static test stand of the Mississippi Test Facility, the first unit to arrive. All earlier S-IC stages were tested in Huntsville, Alabama.

LAUNCH SITE CAPE CANAVERAL

Although in mid-1961, the dimensions of the booster rockets for the Apollo program were not precisely known beyond the Saturn I, there was no doubt about one thing: they would be very, very large—too large for the existing launch facilities at the Cape. The construction of a suitable spaceport was thus one of the first things on which NASA had to decide.

A danger assessment revealed that, for the Saturn C-3, which served as a reference model, the minimum distance between the launch facility and the freely-accessible areas had to be at least three miles. This was one of the many criteria with which NASA, under the leadership of Kurt Debus, in July 1961, began evaluating a total of eight locations for the future spaceport for the lunar missions. In his capacity as NASA Operations Manager, Debus was in charge of all organizational entities involved in launches. The proposed locations were a new spaceport off the east coast of Florida; Merritt Island north of Cape Canaveral; Mayaguana in the Bahamas; Cumberland Island off the southeast coast of Georgia; Brownsville, Texas; the Christmas Islands; Hawaii; and White Sands, New Mexico.

After the first selection round, only White Sands and Merritt Island remained. The main reason for this was that extensive spaceflight infrastructure had been built at both locations in the pre-Apollo period. The fact that launches at White Sands involved flying over a populated area was a clear negative point. Even more serious was the disadvantage created by the sheer size of the individual elements of the launch vehicles. Completely new production facilities would have to be built near the launch site. This would involve the complete infrastructure, from roads to entire cities for the employees, for transporting major components by ship would be impossible in the desert of New Mexico.

Here the Vehicle Assembly Building is seen while under construction, as is Launch Complex 39A in the background. The photo was taken in February 1965.

One factor in Merritt Island's favor was that soon after development of the Saturn C-1 began, the US Army had started construction of a launch facility for this large booster, Complex 34. This facility was ready for use in 1961, and after its completion work began on another facility, Launch Complex 37, further to the north, just south of Merritt Island. Thus Merritt Island was selected, almost inevitably. On August 24, 1961, NASA announced that it planned to purchase 125 square miles of land and construct a space railroad for the moon rockets.

Merritt Island had one great disadvantage, however: mosquitoes. The number of these pests was on a truly apocalyptic scale. Something had to be done, not least for safety reasons for the thousands of workers who would build the tremendous NASA facility. The mosquitoes were present in such numbers that even at 104 degrees in the summer everyone had to wear long shirts and gloves. After a heavy rain, and it rained often there, the salt marshes in the area produced about 15 million new mosquitoes for every 10.75 square feet. And the area concerned was twenty-three square miles in size. The landing rate, as it was ironically and resignedly called at that time, was 500 mosquitoes per minute, per person. In 1962, it was no problem for an entomologist to capture 3.5 pounds of living mosquitoes with a small net. The problem was simply unbelievable, and NASA had to win the war against the troublesome insects before it could begin construction work.

Pesticides were employed as an immediate measure. Brevard County, in which Cape Canaveral is located, provided two spray planes and a helicopter for this purpose. The use of chemicals was not supportable in the long run, however. A more intelligent solution had to be found. And so a system of dikes with inhibition gates was built to the sea. These flooded the mosquito's breeding grounds in summer. The resulting ponds soon formed an El Dorado for minnows, small fish which subsisted mainly on mosquito larvae. Ultimately, an ecological balance took hold, in which the mosquito was no longer the dominant species.

Wernher von Braun now had his new test and development facility in Huntsville, the production

and testing sites in Michoud, and Kurt Debus had his launch facility. All that Debus needed now was a plan for assembling and checking out the huge rockets. He therefore decided on the construction of a huge cubical building, which was called the Vehicle Assembly Building, or VAB for short. There the rockets could be assembled and tested in three work bays. Then they were transported by two huge 3,000-ton tracked vehicles, the Crawler Transporters, to one of the three Mobile Launcher Platforms, each of which weighed 4,100 tons, to the launch complex. Immediately next to the VAB was built the launch control center with its three Firing Rooms. Of the three originally planned launch complexes with the designations 39A, 39B, and 39C, between 3 and 4.3 miles from the assembly building and the launch control center, in the end only two were built. On March 7, 1962, the entire installation was named NASA Launch Operations Center. After the assassination of Pres. Kennedy in November 1963, it was renamed the John F. Kennedy Space Center.

THE LOGISTICS

Not only was the Saturn program an enormous technical undertaking, it was also a logistical challenge of the first order. Components from more than 20,000 contractors and subcontractors not only had to be completed on time, but also had to arrive at the large NASA centers on time so that they could be checked and assembled into larger units. There were things that had never been needed before but now suddenly had to be produced in huge quantities. The fuel load for a single Saturn V, for example, required fifty-six railway cars. The individual stages of the rockets made extended trips on seaways and canals lasting up to seventy days.

To be able to move the Saturn I and later the Saturn V to the transport vessels and test facilities, completely new vehicles were required and for these in turn new, broad and durable transport routes. In Huntsville, for example, an eight-mile-long

This photo was taken on December 17, 1964. It shows the Vehicle Assembly Building and the launch control center at a relatively early stage of construction.

extra-wide road had to be built from the production facilities to the docks on the Tennessee River. In Michoud a new transport route was laid down in such a way that it could utilize the runway of a former airfield.

Each stage had its logistical characteristics. For example, after production in California, the S-II stage was transported by ship to the Mississippi Test Facility (MTF). First it travelled 600 miles by sea to Central America, then through the Panama Canal and into the Gulf of Mexico. The distance for the S-IC stages, which were made in Michoud, was initially only about forty miles. A canal barge took it to the MTF. After tests, the S-II and the S-IC (and previously the S-I and the S-IB) were loaded onto

barges and transported to the Cape in Florida. Logistics for the S-IV and S-IVB were no less complex, although they were somewhat smaller. Here, however, the still large stages had to pass through the heavy Los Angeles traffic to the port, from where ships transported them to the Douglas test facilities near Sacramento.

The Marshall Space Flight Center was soon operating a whole fleet of transport ships. The first was the Palaemon, a 260-foot barge which came from the navy. During a trip from Huntsville to the Kennedy Space Center such a vessel covered 1,500

The barge Palaemon, one of three vessels that sailed the Huntsville–Michoud–Kennedy Space Center route, a total distance of more than 2,100 miles.

miles and took several weeks. From the docks on the Tennessee River it first crossed to the Mississippi, then turned south from Baton Rouge on the Gulf Intracoastal Waterway to St. George Sound, across the Gulf of Mexico to San Carlos Bay, then through the Okeechobee Waterway across Florida to the Atlantic Coast, and then up the Florida Intracoastal Waterway to the Cape Canaveral Barge Canal. Using the intracoastal waterways avoided the open sea as much as possible. Despite this, the ships had to cover 280 miles of open sea. As a precaution, the route was laid down so that the boats were never more than fifty miles from the safety of a port. As a rule they sailed with a crew of twelve. Five formed the ship's crew, six were accompanying technicians and one was mission chief, with overall responsibility for the transport. The trips by these vessels became legendary. One could fill an entire book with curious anecdotes about these journeys. The Palaemon was not alone for long. With the growing need for transports, the barges Promise and Poseidon were also acquired.

As for the transport of the S-IV and S-IVB from the West Coast to Huntsville, they were initially sent on their long journeys by ocean freighter. This would not work for the S-II stage, however, for on the one hand they were very large and on the other required very careful handling. At the end of 1963, therefore, NASA acquired the USNS Point Barrow, a cargo ship dock, from the Navy, which had previously used it in the Arctic. Beginning in 1964,

it transported S-IVB and S-II stages. Later NASA also acquired the Taurus, another high-sea-capable former military freighter. After their arrival in New Orleans they spent another fourteen days on the river barges. They sailed up the Mississippi, then the Ohio, and then into the Tennessee River to Huntsville. The first stage covered 867 miles to Cairo in Illinois and took ten days, where the tugs were changed. Then they sailed forty-seven miles up the Ohio River to Paducah, Kentucky, from where they covered the remaining 323 miles to Huntsville on the Tennessee River.

For the short-distance transports between the production sites in Michoud and the test stands at the maintenance test flight there were two open barges, the Little Lake and the Pearl River. They performed a door to door transport service and could sail right up to the test stands. Each had a thermos container for 30,900 cubic feet of liquid hydrogen. They were stationed at the maintenance test flight, picked up their cold cargo in New Orleans, and thus supported the burn runs of the S-II and the S-IVB at the test stands in Mississippi.

The air transports carried out by the Boeing B-377 PG, on the other hand, were impressively fast. The PG stood for Pregnant Guppy. The aircraft was capable of carrying a complete S-IV stage and

Here a Saturn V S-IC stage on the NASA barge Pearl River being returned from the Mississippi Test Facility to the Michoud assembly facility after successful testing. The photo was taken in August 1968.

At the beginning of 1965, a Saturn I S-IV stage is loaded into Aero Spacelines' Pregnant Guppy at Redstone Airfield, in Huntsville. This view shows that the fuselage could be divided into two separate parts for loading of the voluminous stages.

was frequently used, particularly in the later years of the program. The Pregnant Guppy was a heavily modified Boeing B-377 Stratocruiser, a piston-engined airliner. The first flight of this special transport took place on September 19, 1962. Compared to the original, the PG was several yards longer. Its most noticeable feature, however, was its tremendously enlarged fuselage, which was precisely tailored to the measurements of an S-IV. It got its nickname because it reminded the designers of a pregnant guppy (an aquarium fish).

The company that operated the aircraft was called Aero Spacelines. It cost a great deal of money and effort to get their special aircraft operational

and it almost went bankrupt with this project. Most NASA managers were initially very skeptical. This skepticism at first kept the banks from providing Aero Spacelines with the needed credit. Wernher von Braun and his people were, however, enthusiastic about the idea and supported Aero Spacelines wherever it went, so that in the end the project was successful. The project had a convincing debut with the transport of the S-IV-5. This stage, which was used in mission SA-5, had fallen seriously behind schedule because of testing problems, but now, all of a sudden, it was possible to save three weeks by transporting the stage with the Pregnant Guppy instead of a ship. This aircraft soon became a

Table 1: Saturn Rocket Stage Designations

Type	First Stage	Second Stage	Third Stage
Saturn I Block 1	S-I	-	-
Saturn I Block 2	S-I	S-IV	-
Saturn IB S-IB	S-IB	S-IVB	-
Saturn V	S-IC	S-II	S-IVB

standard part of the US spaceflight program's transportation system. It carried stages of booster rockets, Apollo Command and Service Modules, F-1 engines, the Pegasus satellites for the last three Saturn I flights, and many other oversized NASA transport goods.

The desire soon arose to have a second, even larger transport aircraft and at the same time a replacement in the event that the Pregnant Guppy was not available. The aircraft was put together from components of three other Boeing 377s. The cockpit, the forward fuselage, the wings and the engines came from a Boeing C-97J, a military transport version of the Stratocruiser. With this aircraft it was possible, not only to transport the S-IVB, but also the Lunar Module adapter of the Saturn V. The aircraft was originally supposed to be designated the B-377 VPG, with VPG standing for Very Pregnant Guppy, but finally a better name was found in Super Guppy. The aircraft was not only somewhat larger but also had a better performance, for unlike its predecessor, the Super Guppy was powered by four Pratt & Whitney T-34-P-7WA turbo-prop engines.

S-I AND S-IB

The engineers and technicians in Huntsville had meanwhile made good progress on the concept of the S-I stage. Like most development projects, during the evolution of the rocket between the start of development in 1958, and its maiden flight on October 27, 1961, there were changes, but remarkably few. Soon, however, critics spoke out, questioning the wisdom of the cluster concept. They had obviously very quickly forgotten the emergency situation from which Wernher von Braun was forced to select this concept, but a certain concern was justified. The turbopumps were the Achilles' heel of rocket technology of the time. They worked in the technological limits of what was then possible, and the S-I stage had sixteen of them, two for each engine. Pundits of the day found a connection to American history. The hopeless position of Gen. George A. Custer at the Little Bighorn in battle against the combined forces of the Sioux, Cheyenne, and Arapaho was called "Custer's Last Stand," a dictum that everyone in the USA knew. Critics jokingly referred to the S-I stage as "Cluster's Last Stand."

In fact the design, with its eight engines, nine fuel tanks and the control system which linked everything, was a plumber's nightmare. The affair gained even more in complexity when the S-IV upper stage, with its six additional engines, was added. Fortunately, in this case it had a monolithic tank design.

As few changes flowed into the S-I stage between the start of design and the maiden flight, the more there were during mission operation. This was not due to problems that arose, but to the incremental development that NASA had prescribed from the outset in the early Saturn project. The first four units were designated Block 1. There was only one active first stage, the S-I. They carried only a mockup upper stage. The S-I was also not equipped with aerodynamic control surfaces. With the S-IV, the Block 2 version had an active upper stage. At its base it was equipped with a ring of aerodynamic control surfaces and it had improved H-1 engines.

The S-I stage of the Block 2 Saturn was again somewhat modified, forming the first stage for the Saturn IB, and was designated S-IB. Despite these changes the production and test methods during the entire production of the Saturn I and IB remained almost unchanged. Redstone and Jupiter fuel tanks had been taken as base elements of the S-I and they were lengthened from 40 to 52.5 feet. Their original diameter was retained, 5.8 feet for the

Redstone tanks and 8.75 feet for the central Jupiter tank. As a result the S-I and S-IB could be produced at the existing Redstone and Jupiter production facilities in Huntsville.

The tanks were arranged so that the large oxidizer (liquid oxygen) tank, 8.75 feet in diameter, was in the center, with the eight Redstone tanks grouped around it like the magazine of a Colt revolver. Four of these tanks held fuel, while four contained additional liquid oxygen. The latter had the additional task of absorbing the loads from the upper stage. The fuel tanks, on the other hand, were only responsible for the lateral stiffness of the cluster.

All in all, the Saturn's tanks accommodated 375 tons of fuel. To ensure an even flow of fuel and not unbalance the vehicle during flight, the Saturn I had a complex system of connecting pipes and valves. These saw to it that the level of fuel remained homogenous in all tanks. Each of the four fuel tanks fed two engines but was nevertheless connected to the other units, so that in the event of unequal consumption individual engines did not shut down

before the others. The central oxidizer tank topped up the four outer liquid oxygen tanks, each of which supplied two engines, during the flight.

One of the relatively few development problems concerned the booster's flight stability. Like every large rocket, the Saturn I was also very unstable during the first phase of flight. Its center of gravity was very low, roughly in the middle of the first stage. The center of lift, however, was high in the upper stage. This required advanced flight control processes with constant readjustment of the engines' pivoting mechanism. The amplitude of this control system was, however, very close to the natural frequency of the rocket. Much analytical effort had to be put in to prevent the pivoting motion of the engine from coupling with the natural frequency of the rocket, leading to uncontrollable conditions.

A tricky and time-consuming task was the matter of the heat backflow from the rocket engines. Even for a rocket with one engine it is scarcely possible to determine this for every combination of speed and altitude. The phenomenon is called base

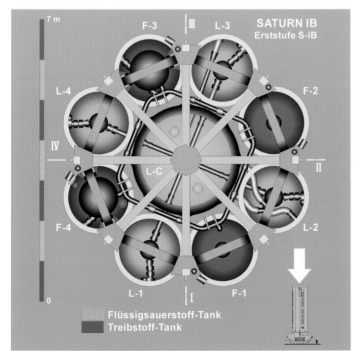

Graphic representation of the tank arrangement in the S-I and S-IB stages.

The completed S-I stage for SA 1 undergoing checkout in Building 4705 of the Marshall Space Flight Center. The photo was taken on January 23, 1961.

heating. The backflow develops when the engine exhaust gases interact with the trailing shock wave. This can result in zones in which the super-hot gases are caught in the vortices at the base of the rocket, similar to vortices at a dam. In the case of the Saturn I, however, the fuel-enriched exhaust gas stream from the turbopumps was also in this area, with the resulting danger of fire and explosion.

The Saturn I cluster with its eight engines was a Pandora's box when it came to base heating. To get an idea of what awaited them, the design team carried out what were called cold flow tests. Using a model of the Saturn I, they conducted them in the wind tunnel of NASA's Lewis Research Center. The model was equipped with eight engines each producing 247 pounds of thrust.

After the developers had clarity about the stream behavior, corrective measures were implemented. The engineers arranged the engines

in a cruciform shape to minimize dead air and zones of turbulence. The four inner engines were placed in the center of the base plate standing close together, the four remaining engines were placed outside at a certain distance from the inner engines. These had to be prevented from getting in the way of the central engines when they pivoted. On the extreme outside were alternating long and short guide plates. The purpose of the short plates was to direct the fast-flowing air in the direction of the central engines so that no dead air areas formed near them.

It had previously been standard procedure for the ABMA to carry out development and production itself, but as the Saturn program grew rapidly this was no longer entirely adhered to by the ABMA (and later NASA), and gradually production of the stages had to be turned over to industry. Even afterward, however, the Marshall Space Flight Center retained its production capabilities for a long time.

The Saturn IB stages S-IB-7, S-IB-9, S-IB-5, and S-IB-6 in the Michoud Assembly Facility, in 1967.

The Saturn I program reflected this transfer process very well. The first eight units of the Saturn I stage were built by Wernher von Braun's organization. The last two S-I stages were manufactured by Chrysler's Space Division as were all later S-IB stages. It was the same with the S-IV upper stage. The prototypes and two of the five production units for the Saturn I program were assembled by NASA. The four other units and then all S-IVB stages were built by Douglas. Incidentally the Chrysler Company had not been selected from five candidates as manufacturer of the first stage if the Saturn I until several weeks after the maiden flight of the Saturn I. One year later the production contract with Chrysler was expanded to include the Saturn IB's S-IB stage.

Essentially the new S-IB first stage retained the shape and size of its predecessor, the Block 2 S-I stage, however the upper section had to be modified to accommodate the S-IVB with its greater diameter and higher weight. The aerodynamic fins were modified to give the necessary stability to the longer and heavier vehicle in the first phase of flight. The engines were arranged in exactly the same way as those of the Saturn I. Successive improvements initially increased the performance of these engines to 200,000 pounds of thrust, while later units produced 205,000 pounds of thrust. It proved possible to save a considerable amount of weight, a total of ten tons, in the S-IB stage. The previous test flights had shown that a consciously conservative way of thinking had caused many components to be designed too large.

Table 2: Large Engines of the Saturn Rockets and Number per Stage

Type	First Stage	Second Stage	Third Stage
Saturn I Block 1	8 x H-1	-	-
Saturn I Block 2	8 x H-1	6 x RL-10A-3	-
Saturn IB	8 x H-1	1 x J-2	-
Saturn V	5 x F-1	5 x J-2	1 x J-2

H-1: MILITARY ENGINE FOR THE FIRST SATURN

As a rule, development of a rocket engine began years before the start of design work on the stage it was to power. This was due to the complexity of propulsion systems and the extensive and lengthy tests that had to be carried out. The engine, and the fuel combination with which it was powered, were always determining factors for the propulsion stage, defining the size and arrangement of the tanks, the entire piping and valve system, the connecting elements between the stage and the engine, and much more. As today, at that time the development of a new engine took about five to seven years. If technological innovations were included, which made accompanying development and demonstrator programs necessary, it could take much longer.

In the case of the development of the Saturn I, it was initially a program with which a test stand model could be created as quickly and cheaply as possible. This led its designers to simply turn to the most powerful operational engine in use at the time, the S-3D, which was employed by both the air force's Thor medium-range missile and its army counterpart, the Jupiter. This engine developed 150,000 pounds of thrust. As a result, development of the S-I stage could begin immediately, which gave the program a flying start.

Several modifications were necessary, however. In a first step, the engine had to be "demilitarized," which resulted in some degree of simplification. As part of a weapons system it needed a complex thrust control system, to ensure that it struck a selected

Eight of these H-1 engines powered the first stage of the Saturn I and Saturn IB. This photo was taken in 1960.

Centering of the four inner engines of the Saturn I was carried out in a special device.

target area. This could now also be done away with. Various technical steps were then taken to increase thrust by a good twenty percent. In its new civilian role the S-3D was now also given a civilian designation and became the H-1.

Although the initial specification for the engine called for a performance of 188,000 pounds of thrust, the first models were limited to 165,000 pounds. This was done to improve reliability and also to keep modification costs low, so that it could be used as soon as possible. In the end the H-1 engine was delivered in four successive performance classes: 165,000 pounds of thrust in the first four flight units of the Saturn I and 188,000 pounds for flight units 5 to 10. In the Saturn IB it produced 200,000 pounds of thrust in flight units SA-201 to SA-205, and from AS-206 the Saturn IB was equipped with engines producing 204,500 pounds of thrust.

On December 31, 1958, less than four months after the development and production contract was signed with Rocketdyne, the first engine was already on the test stand. Tests with the cluster of eight engines began in the spring of 1960. The engine received its preliminary flight clearance in the autumn of the same year.

Development of the more powerful versions of the H-1 followed soon after. The 188,000-pound version received its preliminary flight clearance on September 28, 1962. Approval for development of the flying version for the Saturn IB followed in November 1963. The last engines of the variant with 204,500 pounds of thrust were delivered in 1968. Including all of the development, pre-production, and test models, Rocketdyne had built no less than 322 examples of this rocket engine.

Modification work on the H-1 engine was carried out by Rocketdyne at Canoga Park, California. The company had test stands in the nearby Santa Susana Hills and the first development tests were carried out there. All other testing, especially that involved with the interplay of multiple engines, was carried out at Huntsville. Its facilities included a component test laboratory, a test stand for gas generators and several others for the cluster tests. Production of the rocket engines took place at the Rocketdyne facility in Neosho, Missouri. Not far from the factory there was an air force test stand which Rocketdyne could use. Acceptance tests after production were also carried out there.

Rocketdyne produced the H-1 engine in two slightly different versions. Each unit had fittings for a pivot mechanism welded on the combustion chamber area. The inner engines, which were not used to control the rocket, were not equipped with actuators but were stiffened by struts. The outer engines, however, were equipped with pivot actuators. The exhaust systems for the gas turbines were also different for the inner and outer engines. Because of these differences Rocketdyne gave the inner engines the designation H-1C and the outer engines H-1D.

Production and further development of the H-1 engine up to its final reincarnation from AS-207 was not entirely problem-free, but proceeded as could be expected from a mature initial model. During operational flights with the H-1, a minor problem appeared just once in a total of nineteen launches of the Saturn I and Saturn IB, and that did not decisively affect the mission. One engine shut down prematurely during the launch of SA-6. The remaining seen engines were easily able to compensate for the loss of thrust from one engine with a longer burn time.

F-1: THE MOST POWERFUL ENGINE IN THE WORLD

Rocketdyne had been carrying out studies on behalf of the US Air Force for an engine in the one million pounds thrust category since 1955. In 1958, the air force expanded this contract and ordered test models of the combustion chamber and injection elements. There was no actual requirement for these. The contract was issued in the presumption that such an engine would be required for an as yet unspecified program in the not too distant future, but primarily because of the certain knowledge that it would be at least five years before the first test model of such a monster engine could be brought to the test stand. NASA needed a rocket engine of this performance class much more urgently then the air force. When the respective responsibilities of the air force and the new space agency were laid down when the space authority was founded, responsibility for the F-1 program landed in NASA's lap.

In January 1959, Rocketdyne received a contract from NASA to conduct combustion chamber experiments and in March 1959, it succeeded in achieving 1,000,000 pounds of thrust for 200 milliseconds. The system was still light years from a flight-capable version, however. The first full-scale mockup was presented in May 1960. Tests with the engine's gas generator even began in March 1960. Tests with the first prototypes of the turbopumps followed in November. The first burn tests with the complete combustion chamber, which was close to flight standard, took place on April 6, 1961. The engine ran for several seconds with an output of 1,640,000 pounds of thrust.

So far everything had gone breathtakingly fast, given the size and novelty of the task. Apart from the sheer power—ten times greater than the most powerful rocket engines to date—what the designers had in mind was relatively conservative in nature. It was to be based on the technology that was seen as manageable in 1958. Like the H-1 rocket engine of the Saturn I, RP-1, highly-refined kerosene, was to be used as fuel. Liquid oxygen would also be

ENGINES FOR MANNED FLIGHTS

MODEL	RL-10	H-1	J-2	F-1
THRUST	15,000 LB	188,000 LB	200,000 LB	1,500,000 LB
FUEL	HYDROGEN	KEROSENE	HYDROGEN	KEROSENE

M-P&VE-O BELEW JAN 30,1963 M-CP-D 1404   M-MS-G 89-62 REV A

The Saturn engines in comparison. A contemporary NASA drawing from 1963.

used as the oxidizer. The combustion chamber pressure of 1,160 pounds per square inch was twice the current standard, however. And everything had to become very large, much larger than anything built before. The F-1 engine would be used in a cluster of five power plants, in order to propel the rocket to an altitude of fifty miles and accelerate it to Mach 7.

The sheer size of the engine was the actual problem in development of the F-1. Simply scaling up existing engines would not work. Not only did it have to become very big, it also had to be extremely reliable. After all, in the end it was to power a manned rocket. Development was not trouble-free and no one had expected it to be. One of the most difficult aspects of development was the engine's injector system. Its purpose was to inject fuel and oxidizer into the combustion chambers in the correct quantity and correct mixture. To the layman an injector is a seemingly simple device, which at first glance resembles a large shower head. The many hundred individual injector elements were, however, arranged in a complex arrangement of concentric rings, divided by efficiently arranged stages. Their function was to spray 3.3 tons of RP-1 and liquid oxygen into the chamber per engine, per second.

The liquid oxygen was largely sprayed directly, through the so-called LOX dome, a reservoir above

the combustion chamber, which in turn was constantly topped up with liquid oxygen by the turbopumps. The fuel followed a complicated route. Some of it went into the combustion chamber in convoluted guides into the double-walled combustion chamber to initially act as a cooling medium there. Then it passed through the walls of the combustion chamber head into a circular cooling ring divided by thirty-two accumulators, where it met the main quantity of fuel and with it entered the perforated lower cases of the injection elements. The fuel exited through 3,700 tiny openings in the injection plate and 2,600 other openings, combining with the oxidizer, and went to the combustion chamber as well-mixed, finely-atomized oxygen-fuel mist.

This seemingly so simple but in fact highly-complex structure had to function under very high pressure and enormous heat and produce a stable, even combustion. This did not always happen at the start of development, however. The injector went through the usual development steps of difficult, experimental hardware, ever closer to the production version. It took time to work toward the ideal composition of materials, determine the ideal thickness of the walls and their geometric shape. The first flow tests were conducted with water. Numerous experiments gradually revealed the correct number of injector elements, their arrangement in

F-1 engines are stored in the so-called F-1 Engine Preparation Shop, building 4666 of the Marshall Space Flight Center. The photo was taken in March 1965.

the combustion chamber head plate, the number of holes for the oxidizer per injector, and the number of holes in the case for the fuel. The injection pressure and flow volume were gradually raised. Finally the time came for the first "hot runs."

These proved to be relatively unproblematic. This was due, however, to the fact that the first test engine was run at a maximum thrust of just 1,011,600 pounds. Despite this, in about ten percent of cases spontaneous combustion instabilities occurred. The most unsettling thing was that there was no warning that they were about to occur. They simply happened. They were there from one fraction of a second to the next. At first they didn't cause much damage, but they could potentially ruin the engine. As a safety measure the engineers installed accelerometers on the thrust chamber. As soon as they detected that the engine was beginning to run roughly, it was

immediately shut down. In this way, at least they did not lose the valuable engine; instead only a certain amount of testing time was lost.

The Rocketdyne engineers then began experimenting with a large number of different injector arrangements. To obtain a larger data base, they carried out tests with an experimental transparent low-pressure combustion chamber, a small-scale model of the F-1 combustion chamber. Then they used super-slow-motion films and Schlieren photography to precisely observe the injectors at work. Tests were also made with irregular fuel feed, but nothing produced clear results.

Most promising was a method which had already worked with the H-1 engine: the so-called bomb test. Small explosive charges were set off in the combustion chamber while the engine was running to artificially produce the feared instabilities.

All new designs were now investigated with these methods. In the end the most promising was selected and regular burn experiments were conducted with it. In fact everything seemed to be under control, until during an experiment on June 28, 1962, an oscillation of such a scale built up that an emergency shutdown was unsuccessful. The engine exploded. Quite obviously the problem had not been solved.

NASA now employed an overkill strategy. Project Go was begun and a committee of the most capable authorities in the field of engine technology from across the country was appointed. But the task proved a hard nut, even for this select board, and ten more of these spontaneous burn instabilities occurred during experiments. Two more engines were lost and several more were damaged.

Ultimately, after numerous analyses, thirteen different separator plate designs and fourteen different injector arrangements were tested. Sometimes the improvement process proceeded at a snail's pace. It lasted three years and ran in parallel to "normal" engine development, but in the end the problem was solved. The spontaneous combustion instabilities never reappeared. Until the end, however, the problem was never completely understood. To this day it remains a constant problem in large rocket engines, for which an individual empirical solution has to be found. To the disappointment of the engineers, a model was never found with which they could generally overcome the problem.

The matter of the spontaneous combustion instabilities was in no way the most demanding in the F-1 program. Rocketdyne also had to commit much manpower to development of the turbopumps. They were a technical miracle. The oxidizer pumps,

Here technicians of the Marshall Space Flight Center install the F-1 engines on the thrust structure of an S-IC stage at the test stand for static firing tests in Huntsville. The engines, each weighing ten tons, were only installed once the stage was mounted on the test stand.

Installing an F-1 engine on an S-IC production stage at the Michoud Assembly Facility, June 1969.

fuel pumps, and turbine were mounted on a common driveshaft. Fuel cooled their bearings during operation. The oxidizer pump was capable of moving more than 26,417 gallons of liquid oxygen per minute, while in the same period the fuel pump delivered 15,850 gallons of RP-1 to the engine.

The oxidizer pump supplied the thrust chamber and the gas generator with oxygen. The pressures in the various line systems from the tank to the pumps, from the pumps to the gas generator and to the engine's LOX dome had to pass through a complex system of valves, compressors, and inducers to be able to deliver the liquid fuel to the systems at the correct pressure and temperature. The turbine, which powered the separate fuel pumps, was also an impressive piece of technology. It produced 41 megawatts. The developers placed it on one end of the shaft. On the other end was the fuel pump. Thus the elements which produced the greatest temperature differences were farthest apart from each other.

The turbine ran at a temperature of more than 1,472 degrees Fahrenheit, while the oxidizer pump was at minus 301 degrees.

Development of the turbopump also resulted in a number of incidents, resulting in the loss of a total of eleven pumps. Twice there were structural failures and nine times explosions. Five of these took place during engine tests, four during separate turbopump tests. The causes differed: design errors, materials fatigue, wear. None of these problems were unexpected in the development process for a high-technology system, and all were addressed one by one.

Rocketdyne carried out the development tests with the F-1 engine at nearby Edwards Air Force Base. At the Edwards Rocket Engine Test Site there were five different engine test stands and a multipurpose stand for combustion chamber and injector experiments. The systems in the Mojave Desert had their limitations, however, for the available fuel tanks could not support a burn time of more than twenty seconds duration.

The produced engines, which had to demonstrate much longer running times, were initially sent to the Marshall Space Flight Center in Huntsville and beginning in 1963 the engines thundered through the valley of the Tennessee River. On April 16, 1965, in Huntsville, a test was carried out with the S-IC test stage S-IC-T, with five engines running simultaneously. The test lasted just 6.5 seconds, but for the first time a Saturn V first stage produced 7,418,000 pounds of collective thrust, more than any other rocket stage before it.

On September 6, 1966, the F-1 engine was qualified for manned use. The last acceptance test of an S-IC at Huntsville took place on November 15, 1966, with the stage for Apollo 8 (S-IC-3). All subsequent acceptance tests for completed S-IC stages were carried out at the Mississippi Test Facility. Nevertheless, almost until the end of the Apollo program, individual tests with F-1 rocket engines were occasionally carried out at Huntsville and Edwards.

The combined burn duration for all F-1 engines, including all tests, was seventy-eight hours. That doesn't sound like much, but the burn time per mission was no more than a good 160 seconds. Thirteen times, five engines were used for a flight mission and sixty-five flight engines were constructed. The engine proved 100 percent reliable in all of these flights.

RL-10: FROM THE CENTAUR TO THE S-IV

The contract for the S-IV upper stage was issued to Douglas Aircraft Corporation in April 1960. The propulsion unit, 42.65 feet long, was supposed to be powered by six Pratt & Whitney RL-10-A-3 engines. This was the same rocket engine then being developed for the Atlas Centaur. When design of the Saturn I began, there was not a great deal of experience with hydrogen as fuel. The trust that this new technology could also be used for the Saturn I was primarily based on the assumption that by the time that the engine began use in the Saturn it would already have a long flight history to its credit. It was hoped that it would be beyond its teething troubles by then, but the development of the RL-10 engine proved both difficult and filled with slip-ups. In the end the Centaur program was two years behind schedule. It turned out that the first successful use of an RL-10 engine on an Atlas took place in November 1963, just two months before the engine's first flight on the Saturn.

The Marshall Space Flight Center signed the contract with Pratt & Whitney on August 10, 1960. At that time the engine was still called LR-119. During development the design underwent numerous changes. Along with this went repeated changes in the designation. Finally, at the end of the selection process, the remaining variants, which were to be used both for the Atlas Centaur and the Saturn I's S-IV stage, were designated RL-10-A-3. On March 29, 1961, NASA defined the ultimate configuration of the S-IV stage. It was to be equipped with six of the new engines.

Development of the RL-10 involved 707 burn tests with a total running time of 71,000 seconds until qualification. As was expected with this new technology, it was difficult and time consuming. More than 230 engines were produced for the testing program alone.

The first test ignition of the RL-10 took place in August 1959. Flight certification for use on the Centaur followed two years later. On June 9, 1962, Pratt & Whitney ended flight rating tests for the RL-10-A-3 version. The first hydrogen-oxygen engine in the world had been cleared for production.

41

Integrated into the Centaur stage, it first had to overcome its teething troubles, however, and was a long time before the Centaur achieved a consistent series of successes. Of the first five flights of the Centaur stage on the Atlas, three ended in failure. The first successful use of a Centaur took place on November 27, 1963. That was just two months before the first flight of an S-IV with six RL-10 engines. Despite its difficult beginnings, the RL-10 was a complete success in the Saturn program. A total of thirty-six units were used in six missions by the Saturn I Block 2 series. Later versions of the RL-10 remain in service to this day. As these lines were being written, the latest version of the Centaur upper stage with the 455th RL-10 operational engine placed a US Navy communications satellite into a geostationary transfer orbit.

Production of RL-10 engines.

J-2: THE ENGINE FOR THE MOON

Unlike the limited career—at least in the Saturn program—of the RL-10 rocket engine, the J-2 engine was of vital importance to Apollo, Skylab, and ASTP programs. The J-2 had been added to the program at the recommendation of the Silverstein Committee. What was requested was an engine with at least ten times the performance of the RL-10. At the beginning of 1960, NASA issued invitations to tender for this power plant. Four companies submitted proposals and on June 1, 1960, the Rocketdyne Division of North American was named the winner. The preceding model, the RL-10, was still three and a half years away from its first use in the program.

The contract included a very special passage in a prominent place. Rocketdyne was not only asked to achieve a quantum leap in performance capability for chemical power plants but also ensure maximum safety for manned operations.

The manned flight aspect resulted in the choice of a rather conservative start for the design. The engine's size was challenge enough. The specific impulse of 425 seconds was rather moderate for an engine with this fuel combination. With a combustion chamber pressure of just over 750 pounds per square inch, however, this was not carried too far. And finally the engine was designed on the basis of the gas generator cycle, which at the time was not exactly revolutionary in technical terms. One aspect of the design was very advanced, however. As the J-2 was not only supposed to place the spacecraft into orbit, but also insert it into a lunar transfer trajectory, it had to be designed to be reignitable.

Just six weeks after the signing of the contract, the development team had put together the first experimental J-2 components, namely a full-scale injector head. Then on November 11, 1960, the engineers began the first burn experiments. Other parts of the engine also went into development quickly. At the same time the test stands were whipped into shape and new test facilities were built.

Rocketdyne was fortunate in having its test installations more or less around the corner. Directly above the production facility in Canoga Park, in the northwestern part of Los Angeles, was the Santa Susana Field Laboratory. There was a scattered collection of test stands, control bunkers, and depot barracks there which extended into the small canyons and dry valleys of the Santa Susana Mountains. Before Rocketdyne occupied this area, nearby Hollywood Studios had shot the outside scenes for its westerns there. No longer was the clattering of hoofs, the war cries of Indians, or the banging of Colts to be heard; now these had been replaced by the roaring of rocket engines.

Later trials with an improved version of the J-2 took place at the air force's Arnold Test Center in Tullahoma, Tennessee. Reignition experiments were also carried out there, for the facility had a vacuum test stand which could simulate altitudes of more than sixty miles. And then there were tests at the Marshall Space Flight Center in Huntsville, and of course acceptance tests of the S-II stage in Mississippi.

The first burn tests with the complete engine began in Santa Susana in January 1962. By the end of the summer they had achieved burn times of up to ninety seconds, and on October 4, 1962, the test unit ran for a duration of 250 seconds. The J-2 would have been the first new technology engine whose development proceeded without significant problems, however. The injector was one of the most difficult components of an engine, and this part was no exception when it came to the J-2.

Rocketdyne technicians check out a J-2 engine.

Close-up of the J-2 engines of Apollo 6. They were to cause difficulties during the flight. The photo was taken in the Vehicle Assembly Building during stacking of the S-II stage.

As on the RL-10 engine, on the J-2 it was the task of the injectors to ensure a controlled, stable mixture and burn of the hydrogen and oxygen. The fuel flow rate was considerably greater, however. The J-2 produced 200,000 pounds of thrust, the RL-10 just 15,000. When the engineers at Rocketdyne began developing the injection system, they tried it with a design which had been quite successful in conventional engines, namely with a flat copper injector. To their dismay, however, during the first longer burn tests the engineers saw green flames shooting from the injector and the material simply burned off.

The project management at the Marshall Space Flight Center watched for a while as Rocketdyne struggled with a series of remedial measures and redesigns; but they did no good, and the injectors continued to burn through. Then NASA offered its assistance, for not long before the space authority's Lewis Center and Pratt & Whitney had found a brilliant solution to this very problem during development of the RL-10. To cool the injectors, which injected the fuel, Pratt & Whitney used an arrangement which had its origins in nuclear technology. It consisted of a plate made of superimposed layers of steel wire, which were sintered under heat. This resulted in a highly porous injector which was cooled by hydrogen passing through it in a process called transpiration cooling. Pratt & Whitney called this technique Rigimesh.

The Rocketdyne engineers reacted with what management called the "not invented here" syndrome. This refers to the rejection of solutions that came from outside the company, possibly even from a competitor. In other words: the Rocketdyne project heads refused to accept the solution of a competitor. Finally, the NASA project managers saw themselves forced to make a small example. The Rocketdyne engineers were more or less ordered to the Lewis Center, where they were briefed on the principle of Rigimesh. That helped. Thanks to pressure from NASA and the use of the Rigimesh principle, the problem disappeared and never returned.

There were many other problems, but all were gradually resolved; but one thing loomed over the development of every component. It was a problem that had been anticipated but whose consequences, in all their complicated details, could not be overlooked. It concerned the insulation technology that had to be developed. It became an almost nightmarish project within the project. By then a great deal of experience had been gained in using liquid oxygen as an oxidizer. It was a liquid that had to be handled at temperatures of about minus 292 degrees Fahrenheit, but liquid oxygen was kind to the designers. They could make use of the advantage that every component that came into contact with liquid oxygen began to freeze over. This layer of frost worked as a natural insulator. It worked so well that in the beginning the designers did not have to design many components with insulation.

Liquid hydrogen, 25 degrees colder then liquid oxygen, did not tolerate even the slightest carelessness. If air came into contact with the super-cold liquid hydrogen surface, it became liquid and ran down. There simply was no frost. Consequently not only was the liquid air running down an annoyance, it also warmed the hydrogen in the tanks behind it. Thus all parts of the

J-2 which came into contact with liquid hydrogen had to be insulated. Sometimes simple vacuum tubes sufficed, but sometimes it was a nightmare, especially when the material's creeping properties were appreciated.

Design of the hydrogen feed tubes was a particular challenge. These tubes had to be flexible, because they moved when the engine pivoted. The engine's pivot range was 10.5 degrees. The diameter of the tube was 7.87 inches and it was 20.8 inches long. When the engine pivoted it was compressed or stretched by up to 4.5 inches and that in a twisting, angular movement. Nevertheless the hydrogen had to continue running to the engine undisturbed and in the same quantity. Making this thing absolutely sealed was an immense technical challenge. The ultimate design consisted of a vacuum tube, a sort of double bellows, which was attached to the combustion chamber support ring by harmonica-like

A J-2 engine on the test stand.

actuators. Although it looked quite banal, the thing was ultimately one of the greatest technical achievements of the entire J-2 program.

An effort was made to create as few places as possible that required insulation, but the attachment points could not be made entirely free of joints, tubes, lines, and rods. The danger of a drop of hydrogen with its possible consequences, in the worst case ending in a devastating explosion, existed everywhere. In the end, however, the J-2 still required 112 gaskets of various sizes.

The J-2 received its flight certification in the summer of 1965. The individual engines showed themselves to be unusually long-lived. For example, one was used in 30 successive tests, five of which went beyond the normal flight duration of 470 seconds. The first flight engine, which was allocated

to mission AS-201, had its successful acceptance burn run in Santa Susana on February 26, 1966.

In July 1966, NASA extended its existing production contract with Rocketdyne to the production of a total of 155 J-2 engines, to be delivered by the end of 1968. Part of this contract was also an improved model which developed 230,000 pounds of thrust. Rocketdyne had carried out the necessary work beginning in 1965, and since the spring of 1966, it had been under test at the Marshall Space Flight Center's facilities. The mission planners intended to use this new engine in the Saturn IB from AS 208 and in the second and third stages of the Saturn V from AS 504. In the end deliveries continued past 1968, but when Apollo 11 flew to the moon in July 1969, the last J-2 had been delivered.

UPPER STAGES OF THE SATURN: S-IV AND S-IVB

On January 26, 1960, NASA convened a meeting of suppliers to brief the industry on the requirements for the new upper stage of the Saturn rocket. Wernher von Braun and his people described the plan, handed the representatives of the twenty companies present thick packages of documents, and informed them that they had just one month to make an application to develop and produce the stage. Eleven submissions were received on February 29, and they were subsequently reviewed by a selection committee. The screening ended on April 1, and on May 26, the winner was announced. It was the Douglas Aircraft Company, which had beaten out Convair by a narrow margin.

At the end of January 1962, NASA decided to initially develop the two-stage Saturn I and then a three-stage, significantly more powerful version called the Saturn V. In July, however, the space authority decided to also build an interim version, the Saturn IB. It received an improved version of

the S-I first stage, and as its second propulsion unit the S-IVB, which according to the original plan was not supposed to fly until the Saturn V. The Saturn IB offered the possibility of working out critical flight maneuvers and testing important elements of the Apollo program in space before the Saturn V became available. In the end this made the S-IVB the veteran of the Saturn program.

Most of the design features of the S-IV were included in the S-IVB, although it had a single J-2 engine instead of the cluster of six RL-10 rocket engines. This single engine was twice as powerful as the previous six engines and structural changes were required to accommodate it.

As long as the earth orbit rendezvous method remained the mission mode of choice for the Apollo program, Douglas planned the diameter of the S-IVB stage at 18.35 feet. This mission mode required at least two launches. For this reason the S-IVB was to be capable of waiting in earth orbit for at least

Here technicians prepare the S-IVB-206 for integration with the Skylab 2 Saturn IB. It was the first time that a Saturn IB had been assembled in the Vehicle Assembly Building, which had actually been built for final assembly of the Saturn V.

30 days for the rest of the equipment. This mode would have placed dramatic demands on the insulation technology, to avoid significant quantities of liquid hydrogen from being lost as outgas during this waiting phase.

This changed with the choice of lunar orbit rendezvous. The load increased, because now the entire mission had to be completed with a single rocket, but the waiting period in orbit in active mode

The S-IVB stage of the Saturn AS 506, better known as Apollo 11. Here it is being lifted onto the S-II stage in order to be connected to it.

was reduced significantly. The task of the S-IVB was now to place itself, with the Apollo and Lunar Module, in a parking orbit at about 115 miles above the earth. This would require about one third of its fuel. It would remain there for about two or three orbits of the earth. On reaching the insertion window for the lunar trajectory, the S-IVB would ignite again and place it and its payload on a trajectory for the moon. This required the rest of its fuel. For the new mission mode the diameter of the S-IVB had to be increased to 21.65 feet. In a practical move, this new diameter was adopted for the Saturn IB.

There were six basic systems on board the stage: propulsion, flight control, electrical, instrumentation, and telemetry. A series of individual segments of the S-IV and S-IVB were built in Long Beach and in Santa Monica. Final assembly took place in Huntingdon Beach, California. There each stage passed through a series of production stations, until at the end the J-2 engine was installed. Then the stage was painted, then it was weighed, and finally it was flown to Huntsville or the Cape in the Pregnant Guppy (in the case of the S-IV) or the Super Guppy (in the case of the S-IVB).

One of the many things that caused doubts during the design of the stage was the arrangement of the tanks. The physical features of liquid hydrogen made it necessary to reverse the usual tank arrangement. On board the S-IVB there were on average ninety-six tons of liquid oxygen and twenty tons of liquid hydrogen. The layman would now imagine that the fuel tank would take the lower position and the liquid oxygen the upper position. After all, that is how it was on the first stage of the Saturn V. In fact, here they had to be arranged in the reverse, and there were purely practical reason why. The twenty tons of liquid hydrogen needed a volume of 8,935 cubic feet. The oxygen tank, on the other hand, though much heavier, only required a volume of 2,578 cubic feet. Had the engineers placed the oxygen tank above the hydrogen tank, they would have needed much longer piping. In addition, they could not have fed the piping through the hydrogen tank, because its contents were much colder then the liquid oxygen. And it had to remain that way and not be heated by the "warm" oxygen. It would have required extensive insulation. This could only have been avoided by routing the piping around the hydrogen tank, and this would have made it even longer (and heavier). The reverse configuration, on the other hand, worked quite well. The hydrogen could be carried around the much smaller and more compact oxygen tank with relatively short piping.

The fuel and oxidizer were contained in a common tank structure which was divided into two parts by a bulkhead, which was called the common bulkhead. Thus there were no separate tanks with their own domes and walls. This brought with it weight advantages. These were partially offset, however, by the need for an intertank zone to connect this combined tank into a single structural unit. All in all, it resulted in a roughly twenty percent weight saving compared to the usual method. Douglas developed a double-sided, hemispheric-shaped structure, about two inches thick, with a special aluminum skin on both outer sides and a fiberglass honeycomb structure in the middle. This bulkhead separated the liquid hydrogen (-423° F) on the one side from the liquid oxygen (-277° F) on the other. It also served as the end dome for both tanks and as insulation to prevent the warmer oxygen from heating the colder hydrogen, or the colder hydrogen from causing the oxygen to freeze.

Another problem was the tank insulation itself. Long consideration led Douglas to the realization that internal tank insulation was preferable to external insulation. Then the question of materials came up and it was found that balsa wood would be well-suited and met almost all of the requirements. The problem was that the entire worldwide production of balsa wood would not have sufficed to meet the requirements of quantity production of the S-IVB stages. Douglas had no other choice but to come up with insulation itself.

After a series of tests with different materials and methods the engineers found the solution. They arranged a three-dimensional matrix of fiberglass fibers in a rectangular frame. Then the entire thing was effused with polyurethane foam and hardened. The result was a stable foam block, 7.9 inches high and 11.8 inches long and wide. This could now be cut into tiles and placed on the tank.

The interior of the tank consisted of a structure with wafer-like depressions. This made it necessary to be specially cut to fit precisely into the depressions. The exact shape of each piece—and there were no less than 4,300 pieces—had to be precisely calculated. Each part was given a number and was then placed at a precisely defined place on the tank. A sort of step had to be cut into the edges of these tiles which went over the step of the wafer structure and could be attached to the next tile. The technicians, wearing cleanroom garments, face masks, and shoe covers, climbed through a hole into the tank. They first worked in one half of the tank, beginning at the point furthest from the access hole, gradually working their way back to the hole. Then the tank was turned and the other half was begun.

During this insulation work, precisely-defined environmental conditions had to be maintained. In the next step, a sort of giant balloon was placed inside the tank and inflated. This pressed the insulation firmly into its bedding. At the same time the temperature was raised to 109 degrees Fahrenheit, which better distributed the binder. At the very end a specially-impregnated fiberglass mat was applied and then the whole thing was baked for twenty-four hours at 160 degrees Fahrenheit. Finally everything was carefully cleaned, after which began the installation of other tank components such as filters, valves, measuring instruments, and sensors.

Filling the tanks was a science unto itself. The procedure was essentially the same for the hydrogen and oxygen, only the quantities and filling rates differed. The oxygen tank was first purged with nitrogen, the main purpose of this being to remove any flammable gases from the piping, then the Slow Fill began. The tank was initially filled with liquid oxygen at a maximum rate of 475.5 gallons per minute, until it was five percent full. This served to slowly cool the tank and equipment. Then began the Fast Fill Sequence, in which fuel was added at a flow rate of 951 gallons per minute until the tank was 98% full. Then the remaining two percent was added in the Top Off Sequence. The maximum flow rate during this sequence was 290 gallons per minute. At the end there was the Replenishment Phase with a flow rate of between 0 and 29 gallons per minute, in which the tank was topped up with vaporized oxygen.

Flight control commands came from the ring-shaped Instrument Unit, which sat above the stage in front of the payload adapter. Movements about the yaw and pitch axes were undertaken with the pivoting main engine. Roll control was provided by the thruster system, an arrangement of clusters of four small engines with their own fuel supply in two tanks on the skirts of the stage. During the drift phase in earth orbit and after insertion into the trans-lunar trajectory, this system was responsible for attitude control about all three axes. Immediately before insertion into the lunar transfer trajectory this system also took over forward acceleration.

In flight the stage received its electrical power from silver oxide-zinc batteries. They were housed in the stage's lower skirt together with the auxiliary hydraulic pump. The S-IVB's pyrotechnic system included the detonation cord for stage separation, the propellant for acceleration of the third stage after separation of the S-II stage, and the explosive charges with which the Range Safety Officer could destroy the stage in an emergency.

NASA did not lose a single Saturn during a mission. There were isolated total losses during testing, and development of the S-IV and S-IVB was no exception. On January 24, 1964, a few days before SA 5 was supposed to launch from Cape Canaveral on the first orbital mission of the Apollo program, the pre-countdown for an S-IV test stage took place at the Douglas facility in Sacramento. It exploded and was completely destroyed. Evaluation of the measured data revealed that the oxygen tank had been placed under too much pressure through an error by the test crew, causing the bulkhead between the hydrogen and oxygen tanks to fail. The resulting explosion developed in milliseconds. It was the first time in the world that such a quantity of liquid hydrogen and liquid oxygen had exploded. Previously there had only been theoretical data about the TNT equivalent of such an explosion. The knowledge gained in this accident flowed directly into the future design of test facilities and the necessary safety measures.

Despite this incident, NASA gave the go-ahead for the launch of SA 5 on January 29, 1964. The only change was a directive to closely monitor tank pressure during the countdown at Cape Canaveral. The launch of SA 5 and the following five launches of an S-IV stage were all successful.

There was also one total loss of an S-IVB. This time it was not just a test stage, but a flight model, namely S-IVB-503, which was earmarked for the Apollo 8 mission. On January 20, 1967, the stage was on test stand Beta III in Sacramento for its acceptance test. The countdown proceeded perfectly, and the simulation had already reached T + 150 seconds and was thus just before the simulated separation of the first and second stages. At this point the test had to stopped because of a problem with a recording device. The Douglas crew repaired the recorder, set the countdown clock back, and began the test again. Again all went normally. Eleven seconds before the simulated liftoff, however, the stage exploded without warning. The detonation was huge and all that was left in the area around the test stand was jagged pieces of metal from the stage. The surrounding area was covered and in addition to the Beta III stand the nearby Beta II stand was so badly damaged that it subsequently had to be shut down.

This time the search for the problem took longer. The trail led to one of the helium storage

The just-completed S-IVB stage for Saturn 508 (Apollo 8), seen here in the Vertical Checkout Building at Douglas, where it was readied for transport to the test stand.

spheres, eight of which were arranged in a circle around the J-2 engine. The tank had exploded, fracturing the nearby fuel line. This allowed the hydrogen and oxygen to mix, which led to the explosion of the entire stage. Further analysis revealed that the manufacturer had made the tank out of pure titanium instead of the specified alloy. The tank had previously been tested several times, at extreme over-pressure. In the process it had been "over tested." Because of the material that was used, during the tests the seam gradually weakened until it finally failed during the acceptance test (but not, thank God, during the mission). The results were stricter controls and removal of the manufacturer from the program. Douglas manufactured all subsequent pressure tanks itself.

This is how the artist imagined the first mission by an S-IV stage, in 1965. Depicted here is the moment after stage separation.

51

BASE BLOCK
FOR THE SATURN V: S-IC

When Boeing received the contract for the S-IC stage on December 15, 1961, its design had already been worked out in great detail by Wernher von Braun and his people. Nevertheless there remained many details to be worked out and soon more than 1,000 Boeing workers found themselves in Huntsville to work out the fine details of the design with NASA. At the same time, in Michoud preparations for series production began. At the same time, accompanying technological work, such as wind tunnel studies, took place at Boeing's headquarters in Seattle. Meanwhile, at the Boeing factory in Wichita production equipment was developed and built for use in Michoud, and simultaneously production of components was begun.

The upper dome of the fuel tank of an S-IC stage. The first units were built in Building 4704 of the Manufacturing Engineering Laboratory of the Marshall Space Flight Center.

The first five S-IC stages were built by the Marshall Space Flight Center in Huntsville together with Boeing. They were three test units and the first two flight models. Production facilities designed by Boeing in Wichita were used. The three test units were given the designations S-IC-T, S-IC-S, and S-IC-F. Production of these three units began in 1963. The T-Bird, as it was called, was envisaged for static burn experiments. The S model, which had no engine fitted, was earmarked for structural tests. And the F model, also without an engine, was for compatibility tests with the production facilities, the transport infrastructure, and the installations at the Cape. The first two flight models were the S-IC-1 and the S-IC-2.

After the end of this run-up production in Huntsville, the equipment was sent to Michoud. Construction of a single stage there took about 14 months, nine months for the tanks alone. Of course several stages were always in production at the

The first S-IC fuel tank undergoing tests at the Marshall Space Flight Center, in early 1963.

same time. The first unit, which Boeing itself built in Michoud, was the S-IC-D. It was also purely a test unit with which the engineers and technicians were to gain experience before proceeding to the actual flight hardware. There was also a test model which remained nameless. It also did not consist of original parts, but was made of wood, fiberglass, and other materials. It was needed to test the size and position of newly-designed parts, determine the angle and length of pipes and lines, and assess the optimal placement of wiring bundles.

Chrysler was already at the facility in Michoud and was just producing the last Saturn I and the first Saturn IB. The two companies now had to share the production area. Boeing was given 60% of the space, because the S-IC stages were considerably larger.

The main components of the S-IC were the thrust structures, the fuel tank, the intertank section, the liquid oxygen tank, and the upper structure, which was called the forward skirt.

The thrust structure had to take the loads coming from the engines and simultaneously transmit them to the rocket's structure. It also carried all of the installations and auxiliary equipment necessary for the operation of the F-1 engines and handling the rockets on the ground. For example, there were

four anchors with which the Saturn V was fixed to the ground prior to launch. They were massive aluminum elements, more than twelve feet long and each weighing about 1,320 pounds. Then there were the four engine fairings with their aerodynamic fins, arranged at ninety-degree intervals around the base of the stage. They were made of titanium and had to be able to withstand temperatures of about 2,012 degrees Fahrenheit. On each engine fairing, near the root of the fin, were two solid-fuel retrorockets. Their purpose was to separate the S-IC from the S-II stage and move it away as quickly as possible during stage separation. They each produced 90,000 pounds of thrust for less than a second.

The fuel and oxidizer tanks required special fittings to enable the tremendous flow rate. During filling of the fuel tank this was up to 1,930 gallons of RP-1 per minute. Ten fuel lines, two for each engine, sent 1,295 gallons of fuel per second to the rocket engines. During the countdown the pressure in the fuel tank was maintained by helium from an external source. After liftoff the pressure gas came from helium tanks built into the interior of the oxygen tank. While they were difficult to access there, the raised position was much better, for the super-cold helium tanks could otherwise have frozen the fuel, which was at room temperature.

83 FT.

61 FT.

38 FT.

Taken in November 1967, this photo shows a production step in the manufacture of the S-IC stage at Michoud. Here the fuel tank is being inserted into the thrust structure. In the next stage the intertank section was fitted and then the oxidizer tank was installed.

The gigantic oxygen tank, with its capacity of more than 42,377 cubic feet, was filled in a similar manner to the fuel tank. It began with Slow Fill, about 1,450 gallons per minute. After this phase, which served mainly to pre-cool the tank and components, a veritable torrent of liquid oxygen, 9,510 gallons per minute, began flowing into the tank. The lines which led to the F-1 engines, forty-five feet below, were the very special problem of the oxygen tank. To ensure the required 2,245 gallons per second, no complex lines could wind around the fuel tank. The developers therefore took the straightest route, in the truest sense of the word, and ran the oxygen lines through the middle of the fuel tank.

Unlike the S-IV and S-IVB stages, which had a combined structure for the fuel and oxidizer, the S-IC stage had two separate tanks. This required a connecting element between the two, to make out of them an integrated total stage. This unit was the Intertank, a twenty-three-foot cylinder consisting of massive aluminum elements. It was a supporting structure which had to be capable of withstanding the loads imposed by launch. Its height was determined by the radii of curvature of the two tanks. There was of course a certain amount of empty space in this interstage which was used to accommodate instrumentation, wiring bundles, electrical lines, telemetry transmitters, and many other things.

The Forward Skirt, which was about ten feet high, was placed at the upper end, where the radius of curvature of the oxygen tank began. It, too, was made of a highly stable material. Like the Intertank, it also had many access openings for supply lines of all kinds and many maintenance hatches.

Thousands of tests took place during the stage construction phase and after its completion, involving every component, subassembly, subsystem, and the entire stage. The most important test took place at the end of the process: the static burn of the entire propulsion block. There were two stage test stands. One was in Huntsville, the other in Mississippi. Both were very similar in height and design, but the test stand at the maintenance test flight had two test positions instead of the single one at the Marshall Space Flight Center.

All difficulties were resolved in due time. In 1963 and 1964, there were problems with welding seams and the S-IC-T stage had to be partially scrapped because the oxygen tank could not be saved. In the end, however, this only led to a six-week delay. The dome of the S-IC-S stage's oxygen tank also had to be scrapped. At one point in the program, relatively close to the beginning, production was nineteen weeks behind schedule. This was soon made good, however. Despite the enormous size of the task, the S-IC development, testing, and construction program was notably quiet and problem-free. That was something that could not, by any stretch of the imagination, be said of the S-II stage at North American on the west coast. It was clearly the problem child of the Saturn program.

Taken in October 1968, this photo shows several S-IC stages at various stages of construction at the Michoud Assembly Facility.

THE PROBLEM CHILD: S-II

In April 1960, in Huntsville there was an industry day, during which interested companies could learn about the S-II. At that time it was still envisaged as the second stage for the Saturn C-2 and was to be equipped with four J-2 engines. NASA planned a length of 73.8 feet and a diameter of 21.3 feet. Interest on the part of industry was enormous. No fewer than thirty companies attended the event, but what the company representatives heard must have been sobering, for only seven submitted tenders: Aerojet General, Chrysler, Convair, Douglas, Lockheed, Martin, and North American.

In June 1960, the field was reduced to four, namely Aerojet, Convair, Douglas, and North American.

An event called the Phase II Conference took place at the end of June 1960. There NASA informed the four candidates that they were no longer working on the C-2 version, but were now pursuing the larger C-3 version. The stage's diameter, NASA informed the companies, was being increased from 21.3 to 26.5 feet. In the end the S-II stage had exactly the same diameter as the S-IC, namely 33.1 feet. It was now eighty-two feet long and would be powered by five J-2 engines. With this specification, North American was named the winner of the competition on September 11.

The S-II was a very advanced design for its day. Although its tanks held more than 470 tons of liquid oxygen and liquid hydrogen, the structure comprised just three percent of the fuel weight, even though it was a supporting element. As in the S-IV, they decided on a common bulkhead as a partition between the fuel and oxidizer tanks. This reduced the stage's length by ten feet and saved 4.4 tons of weight.

The stage was assembled in Seal Beach in southern Los Angeles as were most structural elements. Exceptions were the interstage, the upper and lower skirts, and the thrust structure. These components were produced in the North American factory in Tulsa, Oklahoma. Production at Seal Beach was done vertically, which North American saw as the better method compared to horizontal production given the stage's large diameter. The parts warped less and gravity helped precisely align the stage. The "innards" were also installed in the tank in Seal Beach: slosh baffles, instrumentation, cables, pipes,

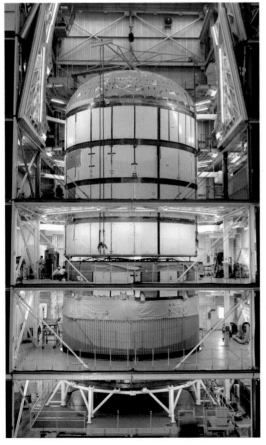

An S-II is assembled in the Vehicle Assembly Building at North American in Seal Beach. The hydrogen tank is being lowered into the common bulkhead with the oxygen tank.

An S-II stage is hoisted into the S-II-A2 test stand at the Mississippi Test Facility (the present-day Stennis Center). All flight stages were tested there. The static firing tests by the test stages took place at the North American facility, in Santa Susana in California.

and sensors. Then the engines were installed and precisely aligned, with intermediate component and subsystem tests, and finally one day the stage was loaded onto the ship and began its long journey into the Gulf of Mexico. Static burn tests then took place at the Mississippi Test Facility. If they were successful, the stage continued to the Cape for launch preparations.

Separation of the S-II stage from the S-IC was very idiosyncratic and was called duel plane separation. Separation of the first and second stages

was begun by the ignition of an explosive guillotine, which separated the two segments from each other. At the same time, the eight braking rockets at the base of the S-IC and the pre-acceleration rockets at the base of the S-II fired. The interstage, weighing 11,464 pounds, was initially only uncoupled, and for about thirty seconds remained connected to the S-II. The separation of this segment did not take place until the engines had achieved ninety percent of maximum thrust. This maneuver demanded great precision to push the 17.7-foot-long interstage downward past the engines. The distance from the J-2 engines, which were running at full power, was just three feet.

This maneuver was an alternative to another option, the "fire in the hole" method. This would have required the ignition of the J-2 engines, the separation of the S-II and S-IC, and the jettisoning of the interstage at almost the same time. The developers did not want to risk this concentration of pyrotechnic activities in the space of just two or three seconds, primarily because of the considerable oscillations caused by the simultaneous operation of the S-IC's retro rockets and the S-II's pre-acceleration engines. They regarded it as reasonable to not carry out the separation until the J-2 engines were running smoothly and would impart no further negative effect on the jettisoning of the first stage.

There were countless difficulties in development of the S-II. One of the most serious was the welding problem. All of the joints had to be as thin as possible (many only .013 of an inch), while having the maximum rigidity and clinical purity. Most of the welds were carried out on aluminum alloy 2014 T6, which was less than ideal but had to be used because of its outstanding characteristics under cryogenic conditions. The material changed its thickness usually every inch or so, for example at places where reinforcements were installed. It suddenly went from half an inch of metal down to wall thicknesses of 0.24 of an inch, and an inch farther back up to 0.62 of an inch. Making a three-foot, high-quality joint there under these conditions was a challenge. But the welds had to go all the way around the stage, more than ninety-eight feet. The slightest temperature deviations and variations in humidity led to microscopic cracks. With a highly viscous material, which liquid hydrogen is, this could have catastrophic consequences. Completely new welding techniques had to be developed and completely new environmental conditions created, to be able to successfully carry out the welds. Despite every precaution, there were repeated undesired incidents. On October 28, 1964, for example, a completed bulkhead on the S-II-S stage cracked during a hydrostatic test, even though a lower pressure was used than what the specification prescribed. The failure was traced to an earlier repair weld, which had been carried out manually at a maintenance opening.

The liquid oxygen tank, with its diameter of thirty-three feet and height of twenty-two feet, was elliptical in shape and had practically no vertical side walls. The dome of the oxygen tank was also the common partition between the oxygen tank and the much larger hydrogen tank above it. The oxygen tank was formed from a dozen wedge-shaped elements, which were attached to "dollar sections" above and below. An exotic new production method had to be developed for the individual elements with their complex curvatures: explosive forming under water. The individual elements were literally blown into shape. A new facility with a 7,450-gallon tank had to be created in Los Angeles. Each element required three explosive processes before it was completely formed.

Production of the oxygen tank dome, which was also the partition to the hydrogen tank, required a precise choreography of successive operations. The first was application of an insulating honeycomb layer of phenolic resin. It formed the bottom of the partition and provided effective insulation. This insulation was then hardened in an autoclave. Production of the honeycomb structure was an extremely difficult task, for the thickness of the material ranged from about a tenth of an inch to up to five inches at the dome cupola. On this base the tank bottom of the hydrogen tank was produced as a mirror image of the oxygen tank dome beneath it. The same explosive forming method was used to create the individual wedge-shaped elements.

Countless tests and fine finishing work were necessary to shape the base of the hydrogen tank, create a perfect surface and ensure that there was not even the smallest gap in the insulation and the individual components. The rest of the hydrogen tank was produced in six individual aluminum rings. The first was twenty-seven inches high, and each of the others 94.5 inches. These rings were then assembled quite conventionally, being bolted together. This made the tank stable, and that was also necessary, for the complete unit was part of the stage's load-bearing structure. The upper dome was made in a similar way to the half shells of the oxygen tank.

Another significant problem was insulation of the hydrogen tank. It took a long while to come up with a usable solution. Unlike Convair with the S-IV stage, North American decided to use external insulation. The original plan, to glue panels of insulation to the outside of the tank, did not work satisfactorily. After numerous experiments, in the end a solution was found that was both surprisingly simple and effective: the isolation was simply sprayed in liquid form onto the tank, allowed to dry and it was simply ground and cut to the correct shape.

Because of all these problems, development of the S-II fell further and further behind schedule. At the start of the program North American planned to carry out its development and flight qualification tests using the experimental stages. They were designated S-II-F, S-II-T, S-II-D, and S-II-S. To make up for lost time and at least get the first flight stage—the S-II-1—ready for the first Saturn V launch, it was decided to shorten the test program. Production of the S-II-D, originally envisaged for experiments on the dynamic test stand, was cancelled. Instead, these tests would now be carried out with the S-II-S, which was subsequently designated S-II-S/D. With this step the S-II-1 moved up one position in production. This was a bad decision, however, as would soon become clear.

During a fueling test at Seal Beach on September 29, 1965, the S-II-S/D was completely destroyed. The investigating team, with the dramatic name Catastrophic Failure Evaluation team, found that the tank had exploded under a load of 144% of the nominal value. It was then decided to use the S-II-T for experiments on the dynamic test stand, however this stage was still at the maintenance test flight facility undergoing static burn tests and was temporarily unavailable. Further delays piled up.

The arrangement of the J-2 engines in the S-II stage can be seen here.

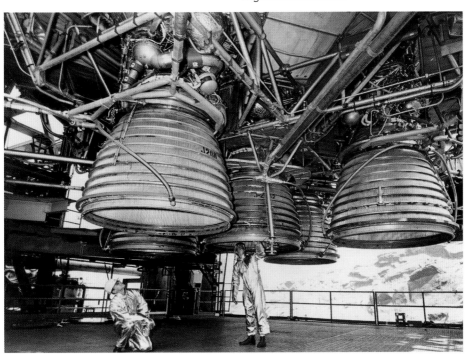

Under pressure from NASA, North American changed the management of the stage development and hired former Air Force Gen. Robert Greer. By the spring of 1966, Greer had mobilized his troops and began an intensive ground test program at the NASA test facilities in Mississippi. On April 23, 1966, for the first time the five Rocketdyne J-2 engines ran together for 15 seconds. The change in leadership seemed to be bearing fruit, then misfortune struck again.

During burn tests on May 10, 11, and 16, each time the engines shut down too soon. This was traced to an error in instrumentation. During two further tests on May 17 and 20, the engines fired for 150 and 350 seconds. During the next planned 350-second test, on May 25, a fire broke out and the test had to be halted. Three days after that, just as the S-II-T was being removed from the test stand, the hydrogen tank and with it the entire stage exploded. Five people were injured and the test stand sustained considerable damage. And the cause? The technicians had carried out pressure tests on the stage, using helium, to determine the cause of the fire three days earlier. They were unaware, however, that another team of technicians had deactivated the pressure sensors and pressure relief valves. As a result, the tank was pressurized beyond its design limit and ruptured.

Not until the beginning of 1969, by which time the last five units were being delivered, did North American really have production of the S-II stage in hand.

May 25, 1966, was nevertheless an important day for the Saturn V. Two states away, in Florida, the full-size Saturn V mockup, designated SA-500, was rolled out. Not only was the enormous model exactly the same size as a real Saturn V, 364 feet, at 216 tons it also weighed the same as a Saturn V without fuel. Three huge tracked vehicles, which NASA had developed and built to transport the rockets, drove it to Launch Pad 39A to be used in tests of the launch facilities.

The S-II stage of SA 506 (Apollo 11) is attached to the S-IC stage in the Vehicle Assembly Building at the Kennedy Space Center.

THE INSTRUMENT UNIT: THE BRAIN OF THE SATURN

The Instrument Unit, or IU for short, bore the responsibility for ensuring that everything that had been created in millions of man hours reached its objective. On the Saturn V this system was a thin, circular structure, less than three feet high, with a diameter of twenty-five feet. It sat between the S-IVB stage and the payload adaptor. Also in this ring were the computer, the gyroscopic system for the inertial navigation system, and various black boxes, in which vast amounts of information was processed and turned into commands for the rocket's control system. The heart of this system was the ST-124 (ST = Stable Table) inertial platform and the Launch Vehicle Digital Computer.

On the Block 1 Saturns, with their dummy upper stages, the flight control equipment was scattered in various places in the adapter area above the stage. These individual boxes contained telemetry transmitters, trajectory tracking electronics, signal processors, and other components. Below them was also the ST-90, the predecessor of the ST-124 inertial platform.

From Mission SA 5, the ST-124 and an improved IBM computer were on board. Then, from flight to flight, they were increasingly integrated into the rocket control system. SA 5 was also the first mission in which all of the electronics, including the flight control equipment, were contained in a cylindrical section above the stage and functioned as one of the booster's independent systems. By this phase it was already called the Instrument Unit. In the Block 2 Saturn it was a ring five feet high, from which four cylinders, spaced at ninety-degree intervals went to the center, where all four cylinders were then joined together. This new element could be installed as a complete unit for the first time. This made launch preparations more flexible and enabled modifications to be made between launches. The four cylinder segments contained the ST-90 platform, the ST-124, the telemetry equipment, and the energy and control package. The cylinders were pressure ventilated and contained an environmental control system to heat the units.

From the mission of SA 9, the Saturn-type cylinder ring was introduced, with the instruments on the walls of the cylinder. The containers were no longer pressure-ventilated and thus the equipment was exposed to the vacuum. Exhaust heat was discharged by radiators.

The ST-124 inertial guidance platform was the heart of the Instrument Unit.

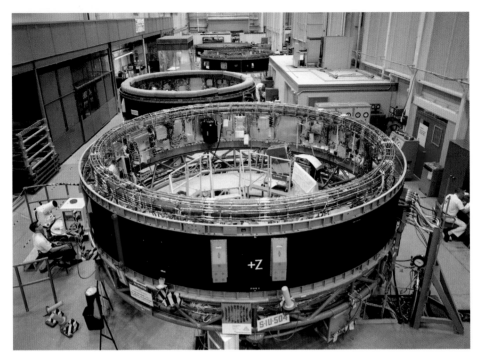

The Instrument Units (IU) of the Saturn IB and Saturn V were made by IBM, in Huntsville. They weighed about two tons.

This and other simplifications meant that not only was the ring lighter, its height was reduced by half.

In 1964, responsibility for construction was passed to IBM. Until then development and production of the IU had been carried out by the Marshall Space Flight Center itself. The IU for the Saturn IB was in principle the same model as for the Saturn V. The idea was to overcome teething troubles during the Saturn IB flights and then hand over a completely developed product for the Saturn V.

The most important functions of the Instrument Unit were flight guidance and control, the control of all flight sequences, management of stage separation, and transmission of flight data and other information to earth. Interestingly the IU was a load-bearing structural component within the rocket. Three stages were below it, and above it were the Lunar Module and the Apollo spacecraft with the three astronauts.

The IU's key elements were the ST-124 inertial guidance platform, the digital computer, and the data adapter. With its three gyroscopes, the ST-124 was a miracle of technology for its day. It had a diameter of 20.85 inches, weighed 115 pounds and was made largely of beryllium. It was made by the Bendix Corporation's Navigation and Control Division.

The coordinate system was fixed just prior to launch and this included a special procedure, with an accurate theodolite situated not far from the launch pad played the most important role. This theodolite sent a beam of light to a small opening in the Instrument Unit, from where it was redirected to the inertial platform. An optical device, an arrangement of prisms, was adjusted so that the beam of light returned to the theodolite. This calibrated the optics, which in turn aligned the inertial platform precisely with the launch azimuth. Receipt of these light signals sent back to the theodolite by the prisms was a key checkpoint in the final countdown of every Saturn V.

Interestingly all of the carefully worked-out autonomous control and correction modes were not needed at all while the first stage was operating. Although the ST-124 unit began operating even before liftoff, and continuously supplied the computer with velocity and flight data, in the first two and a half minutes of the ascent it made no control inputs to the stage. During this phase the rocket was subjected to considerable environmental stress. While flying through the dense atmosphere it had to absorb great dynamic loads, plus wind shear and jet streams. Additional interference caused by control inputs and compensating maneuvers were therefore avoided. During the first phase of flight the rocket therefore flew a preprogrammed flight path entered into the computer. All trajectory deviations, were, however, recorded and the saved data was then used to correct flight path deviations during the second and third stage burn phases.

The computer's processor and memory were unusually advanced for that time, but yet its performance still sounds grotesque to us. The computer and the so-called data adapter subsystem in the Saturn IB weighed 210 pounds, while the Saturn V's tipped the scales at 247 pounds, and that despite its lightweight construction using a magnesium-lithium alloy. Both systems together took up more than 35 cubic feet. The computer was capable of carrying out 3,200 (Saturn IB) or 9,600 (Saturn V) arithmetic operations per second and had a working memory of 100 kilobits (Saturn IB) or 460 kilobits (Saturn V). It is certain that no more elegant programs have ever been written than the flight control software from the Apollo days. Each single bit

of program code had to be precisely considered and every excess bit had to be eliminated immediately.

The independent life of a Saturn V began three seconds after ignition and five seconds before liftoff. At that point the IU took over the rocket. The booster's control system initially carried out a series of preprogrammed maneuvers. These included turning the rocket several degrees away from the launch tower immediately after liftoff to avoid damage to the launch installation. After twelve seconds of vertical flight the roll program was executed, placing the booster on the envisaged trajectory. As soon as the sensors in the tank informed the IU that the fuel level had reached a certain point, it began the command sequence for shutting down the engine and stage separation. After the S-II stage was ignited, the concept of adaptive trajectory control was carried out. Every two seconds the computer compared the present situation and flight parameters with the desired situation at the end of the powered flight of each stage. If it was necessary, the computer gave control inputs to the analogue flight control unit, which then controlled the pivoting elements of the engines or the changed mixture ratio of oxidizer

Here the IU of SA 501 is just being attached to the S-IVB stage. The photo was taken on June 19, 1967.

and fuel. The two flight phases of the S-IVB upper stage were controlled in a similar manner.

If it had become necessary to abort a Saturn mission, then the Propellant Dispersion System, PDS for short, would have been activated. This was a rather euphemistic designation, for it was nothing more than an arrangement of pyrotechnic elements to blow up the rocket if it veered off course. This of course would only take place after the crew had reached safety using the LES, or Launch Escape System. The purpose of the PDS was to prevent Merritt Island, Cape Canaveral, or Cocoa Beach from being laid waste in the event of a Saturn going off course. A special receiver received the coded information, decoded it, checked it, and after all the engines had been shut down a series of explosive charges at strategic positions would have torn the rocket apart. Fortunately the PDS never had to be used.

On the Saturn rockets there were a considerable number of other powerful engines with a short burn duration. Their purpose was to separate the stages from each other, or provide continuous thrust during the brief transition phase between one stage's end of burn and the ignition of the next propulsion unit. The fuel could not be allowed to become weightless, for otherwise the danger would exist that bubbles might form in the lines. It was one of the IU's many tasks to arm and fire these thrusters.

Only solid-fuel thrusters were used for this purpose in the lower stages. Only the S-IVB had a liquid-fuel system which operated on hydrazine, in addition to two solid-fuel thrusters (which were jettisoned after completion of burn). Altogether, there were more solid-fuel rockets in the Saturn booster family than liquid-fuel ones. The Saturn I had thirty-two thrusters of various types, the Saturn IB thirty-one, and the Saturn V twenty-two.

FIRST FLIGHT AND TEETHING TROUBLES

At seven in the morning on November 9, 1967, Merritt Island and nearby Cape Canaveral experienced for the first time an event that would characterize the coming decade. It began with the high-pitched screech of the turbines as they quickly ran up and then it turned into an increasingly loud low-frequency crackling and rumbling, which within a few seconds rose to a roar and rattle and fifteen seconds later sent physically-perceptible pressure waves to the observers at the press stand, more than three miles away. Noises and perceptions like this were not unusual in this area. In the 1960s, several dozen launches were made from there into earth orbit and to the depths of the Solar System. But when the five F-1 engines of Saturn V SA 501 ignited the physical manifestations were such that the question arose—had the Saturn V really launched, or had Florida simply sunk?

Apollo 4 was on its way into orbit. In addition to testing the booster rocket, the mission's purpose was to confirm the structural integrity and compatibility of the combination of booster rocket and spacecraft. The Saturn V's job was to transport the Command and Service Module into a high elliptical orbit from where, in a "power dive." It would accelerate downwards toward the Hawaii archipelago in the Central Pacific from a height of 11,185 miles. This would accurately simulate the ship returning directly from the moon. The spacecraft also included a mockup of the lunar lander.

The flight went almost precisely according to plan. The mission was a decisive step forward for the Apollo program. Its significance could not be overestimated, primarily because of the number of firsts it achieved. It was the first flight of the first and second stages of the Saturn V. Only the third stage had already been tested, for it had formed the second stage of the Saturn IB. It was, however, the first time that the stage had been reignited in orbit. The fact that everything had gone so well and that there had been so few problems gave NASA new self-confidence. The Apollo program was on its way to the moon. Nevertheless, at that time no one would have dared to hope that the first man would walk on the earth's satellite just eighteen months and six more Saturn launches later.

Not all of the problems had been eliminated, however. The Saturn V also had to go through its teething troubles, as its second launch demonstrated. The conditions under which they appeared could not have been recreated during tests on the ground. AS 502 took off on the Apollo 6 mission on April 4, 1968, a few minutes after seven o'clock Eastern Time. The plan envisaged that the S-IVB stage would put the Apollo and the dummy Lunar Module onto a lunar transfer trajectory. Then the Command and Service Module would separate from the stage and fire its engine to break off the lunar transfer trajectory. The object was to simulate a so-called direct-return abort, meaning a flight abort immediately following the trans-lunar injection.

The first two minutes of the flight went almost normally. The engines spewed flames as they were expected to do, but that was almost the last normal thing that the rocket did, for suddenly thrust oscillations began which entered into resonance with the fuel lines and the rocket structure: the feared pogo effect. The sensors in the command capsule measured vertical shocks of +/-0.6 g. This was more than twice the allowable upper limit. At 133 seconds after the rocket left the launch pad parts of the adapter structure, which housed the lunar test vehicle and sat above the Apollo spacecraft, broke off; but despite the hefty vibrations the first stage did its job, the F-1 engines were unaffected and the badly damaged elements stayed together.

The second stage's five J-2 engines ignited two minutes and thirty seconds after the rocket left the launch pad. Seventy-five seconds later engine Number Two registered a loss of power. Five minutes and nineteen seconds into the mission the output

SA 501 (Apollo 4) about eighty seconds after leaving the launch pad.

of the rocket engine again fell off drastically and six minutes and forty-two seconds after liftoff the engine shut down completely, almost two minutes prior to normal burn termination. Just two seconds later engine Number Three also failed. Now just three of the five J-2 engines were still running. In an attempt by the Instrument Unit to compensate for the two engines which had shut down early, the remaining three engines ran for fifty-eight seconds longer than planned. Then the fuel was exhausted, but the speed that had been lost had not yet been made up. The minutes-long loss of power by engine two until it shut down completely, and the subsequent dragging of the dead weight of the two silenced engines took its toll.

Then the J-2 engine of the third stage ignited. Instead of the planned 165 seconds, it had to run for 194 seconds to achieve a parking orbit. This was successful, although there was a considerable deviation from the planned orbit. Just three hours later, after two orbits of the earth, the TLI (Trans-Lunar Injection) maneuver was supposed to take place and accelerate the third stage with the Apollo spacecraft and the lunar test article to trans-lunar velocity.

The increased fuel consumption so far and the changed flight path could easily have been rectified. After the second orbit Flight Director Clifford Charlesworth transmitted the ignition command to the stage, but it refused to carry out the command. Ignition did not take place, stranding the entire combination in a low earth orbit. There was nothing to do but separate the Command and Service Module from the useless S-IVB and let the Apollo carry on the mission alone. The planned trans-lunar trajectory had to be abandoned.

Each of the engines had been tested seven times on the ground. There had been five individual tests for each J-2 and two burns with the entire stage. So how could something like this happen? A solution was soon found for the pogo resonances in the first stage. The failure of the J-2 engines in the S-II and of the S-IVB stage was a mystery, however.

Then something caught the eye of the engineers which previously had been paid little attention. During tests at the Arnold Engineering Development Center in Tennessee the engineers noticed that a thick layer of frost formed on the fuel lines to the spark igniters when the engines were fired at ambient temperatures. The spark igniters were basically also little engines. Their fuel was diverted from the oxidizer and fuel main lines leading to each engine. They not only ignited the engines at the start of the burn phase, but also remained in continuous operation. During ground tests the layer of ice dampened vibrations. In space, however, there was no moisture

SA 502 (Apollo 6) about fifty seconds after liftoff.

which could lead to the formation of ice. There was a bellows on the ignition system. Tests in a vacuum revealed that the bellows began swinging wildly and soon after reaching peak flow loads. The feeds were therefore strengthened and the bellows was replaced by steel lines. The design was the same on both stages, and modifications could be made to both the S-II and S-IVB stages. The problem never reappeared and the way to the moon was clear.

EPILOGUE FOR THE SATURN

The Apollo program, with the development and construction of the Saturn rockets as a significant element, was possibly the most important engineering feat in the history of humanity. A contemporary estimate said that this project, which passed its zenith in 1969, involved 20,000 companies of all sizes, with about 300,000 people taking part. Entirely new industries had been created, significant branches of industry made the biggest advances in their history and completely new products were created, whose affects can be felt in everyday life to this day.

One of the most noteworthy aspects of the Saturn program was its success rate. Even in early press reports it was declared openly that, because of the complexity and new technology, it was expected that half of the first ten Saturn missions would fail. Unbelievably, in the end not one of the thirty-two Saturn missions ended in failure (although SA 502 came closest). The contributions made by the Saturn IB and Saturn V to the Skylab program, which followed the Apollo moon program, are almost forgotten today. And finally a Saturn again entered the limelight at the cusp of a new age in space travel, when it took part in the ASTP Program, the first harbinger of cooperation between east and west in manned spaceflight.

One point is and remains disappointing, however, even more than four decades after the last flight of a Saturn rocket: when the program was cancelled for political reasons, there existed completed hardware worth several billion dollars. Nothing remains but five S-IB stages, two S-IVB stages for the Saturn IB and two more for the Saturn V, two each of the S-IC and S-II stages, four Apollo Command and Service Modules, and four Lunar Modules.

Thus at least two more moon missions and two to three Skylab flights could have been equipped and carried out. It would only have required about five percent of the construction costs already laid out to fly these missions, but ultimately, the USA lacked the will.

Table 3: Surviving Saturn and Apollo Hardware

Stage/ Spacecraft	Present Whereabouts/Original Purpose
S-IB-209	Kennedy Space Center Rocket Garden
S-IB-211	Alabama Welcome Center, Ardmore
S-IB-212	scrapped
S-IB-213	scrapped
S-IB-214	scrapped
S-IVB-209	Kennedy Space Center, Rocket Garden
S-IVB-211	Space & Rocket Center, Huntsville
S-IC-14	Johnson Space Center (was envisaged for Apollo 18)
S-IC-15	Michoud Assembly Facility (Apollo 19/Backup Skylab)
S-II-14	Johnson Space Center (was envisaged for Apollo 8)
S-II-15	Kennedy Space Center (Apollo 19/Backup for Skylab)
S-IVB-513	Johnson Space Center (envisaged for Apollo 18)
S-IVB-514	Kennedy Space Center (envisaged for Apollo 19)
CSM 010	Space & Rocket Center, Huntsville
CSM 105	National Air & Space Museum Washington
CSM 115	Johnson Space Center Houston
CSM 119	Kennedy Space Center (Skylab Rescue/ASTP backup)
LM 9	Kennedy Space Center (envisaged for Apollo 15)
LM 13	Cradle of Aviation Museum, New York (envisaged for Apollo 18)
LM 14	Franklin Institute, Philadelphia (envisaged for Apollo 19)
LM 15	scrapped

SPECIFICATIONS, DIMENSIONS, DIAGRAMS

Saturn I Block 2 (SA 5)	
Country of Origin	USA
Manufacturer	NASA MSFC
Length	164.3 feet
Diameter over Fins	40.7 feet
Total Weight	550 tons
Launch Thrust	1,504,000 pounds
Payload in LEO	23,369 pounds
First Stage: NASA MSFC-S-I	
Power Plant	8 x Rocketdyne H-1
Thrust at Sea Level	8 x 188,000 pounds
Thrust in a Vacuum	8 x 213,120 pounds
Specific Impulse Sea Level	255 seconds
Specific Impulse Vacuum	289 seconds
Burn Duration	147 seconds
Fuel	Kerosene
Oxidizer	Oxygen
Weight Empty	117,500 pounds
Weight Fueled	959,010 pounds
Length with Adapter	95.5 feet
Diameter	21.4 feet
Second Stage: NASA MSFC S-IV	
Power Plant	6 x RL-10A-3
Thrust (Total)	90,000 pounds
Specific Impulse	421 seconds
Burn Duration	480 seconds
Fuel	Hydrogen
Oxidizer	Oxygen
Weight Empty	14,330 pounds
Weight Fueled	114,420 pounds
Length	41 feet
Diameter	18.3
Payload Fairing	
Length	27.9 feet
Diameter	12.8 feet
Operational History (Saturn Block I and Block II)	
First (Suborbital) Launch	October 27, 1961
Launches (Total)	10
Number Successful	10
Last Launch	July 30, 1965

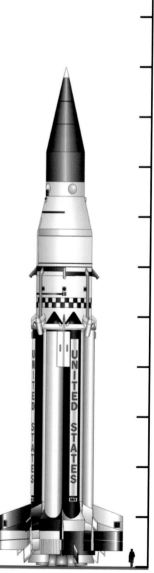

Saturn I / Block 2 (SA-5)

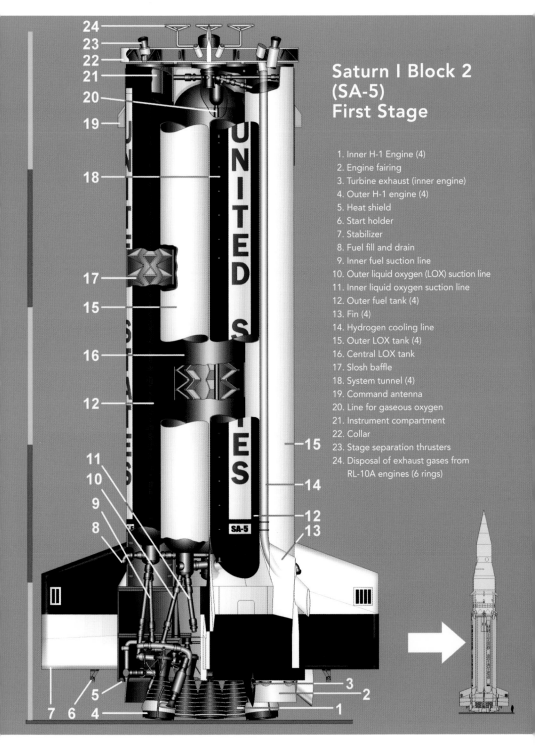

Saturn I Block 2 (SA-5) First Stage

1. Inner H-1 Engine (4)
2. Engine fairing
3. Turbine exhaust (inner engine)
4. Outer H-1 engine (4)
5. Heat shield
6. Start holder
7. Stabilizer
8. Fuel fill and drain
9. Inner fuel suction line
10. Outer liquid oxygen (LOX) suction line
11. Inner liquid oxygen suction line
12. Outer fuel tank (4)
13. Fin (4)
14. Hydrogen cooling line
15. Outer LOX tank (4)
16. Central LOX tank
17. Slosh baffle
18. System tunnel (4)
19. Command antenna
20. Line for gaseous oxygen
21. Instrument compartment
22. Collar
23. Stage separation thrusters
24. Disposal of exhaust gases from RL-10A engines (6 rings)

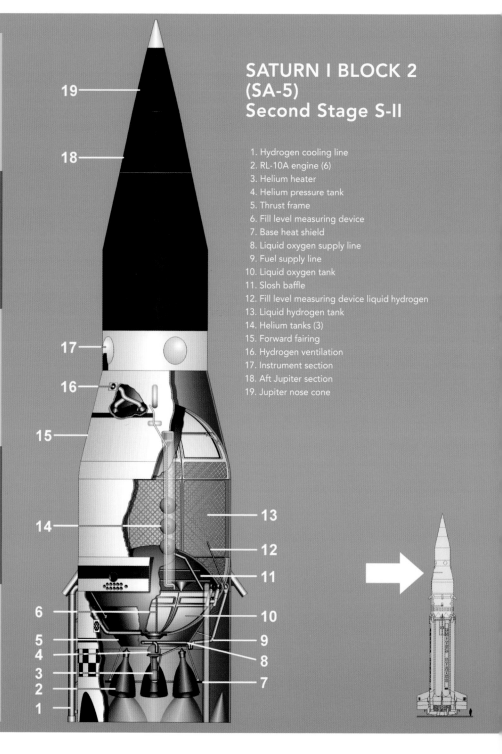

SATURN I BLOCK 2 (SA-5)
Second Stage S-II

1. Hydrogen cooling line
2. RL-10A engine (6)
3. Helium heater
4. Helium pressure tank
5. Thrust frame
6. Fill level measuring device
7. Base heat shield
8. Liquid oxygen supply line
9. Fuel supply line
10. Liquid oxygen tank
11. Slosh baffle
12. Fill level measuring device liquid hydrogen
13. Liquid hydrogen tank
14. Helium tanks (3)
15. Forward fairing
16. Hydrogen ventilation
17. Instrument section
18. Aft Jupiter section
19. Jupiter nose cone

Saturn IB SA 205 (Apollo 7)	
Country of Origin	USA
Manufacturer	NASA MSFC
Length	223 feet
Diameter over Fins	40.7 feet
Total Weight	650 tons
Launch Thrust	1,600,000 pounds
Payload in LEO	48,500 pounds
First Stage: Chrysler S-IB	
Power Plant	8 x Rocketdyne H-1
Thrust at Sea Level	8 x 200,080 pounds
Thrust in a Vacuum	8 x 231,553 pounds
Specific Impulse Sea Level	262 seconds
Specific Impulse Vacuum	296 seconds
Burn Duration	155 seconds
Fuel/Oxidizer	Kerosene/Oxygen
Weight Empty	91,712 pounds
Weight Fueled	989,104 pounds
Length	80.4 feet
Diameter	21.4 feet
Second Stage: Douglas S-IVB	
Power Plant	1 x Rocketdyne J-2
Thrust	231,553 pounds
Specific Impulse	421 seconds
Burn Duration	475 seconds
Fuel/Oxidizer	Hydrogen/Oxygen
Weight Empty	28,440 pounds
Weight Fueled	261,909 pounds
Length/Diameter	14.75/12.8 feet
Apollo Service Module	
Power Plant	1 x Aerojet SPS
Thrust	22,000 pounds
Specific Impulse	301 seconds
Treibstoff/Oxidator	Aerozin 50/N2O4
Weight Empty/Fueled	11,155/13,360 pounds
Length/Diameter	14.75/12.8 feet
Adapter and Escape Tower (LES)	
Length Adapter and LES	81.7 feet
Diameter	12.8 - 21.4 feet
Weight Adapter and LES	12,920 pounds
Operational History	
First (Suborbital) Launch	Feb. 26, 1966
Launches (Total)	9
Number Successful	9
Last Launch	1July 15, 1975

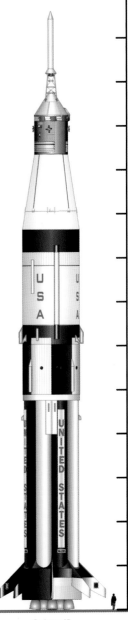

Saturn IB
(SA-205)

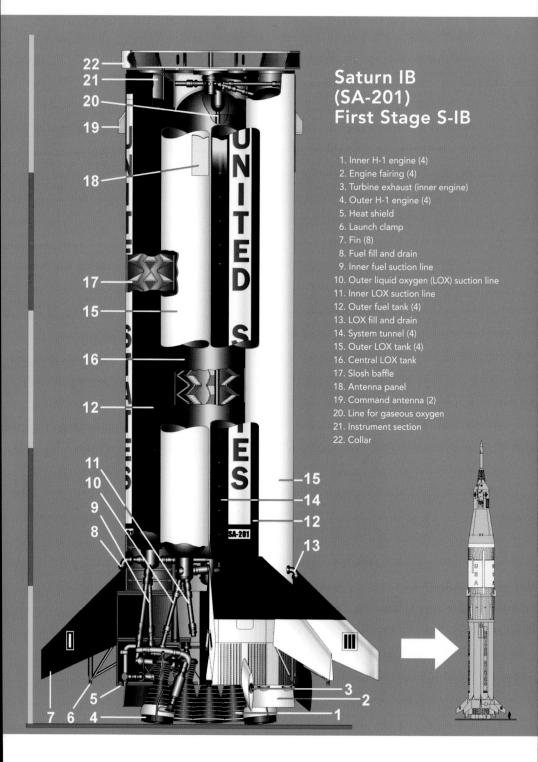

Saturn IB (SA-201) First Stage S-IB

1. Inner H-1 engine (4)
2. Engine fairing (4)
3. Turbine exhaust (inner engine)
4. Outer H-1 engine (4)
5. Heat shield
6. Launch clamp
7. Fin (8)
8. Fuel fill and drain
9. Inner fuel suction line
10. Outer liquid oxygen (LOX) suction line
11. Inner LOX suction line
12. Outer fuel tank (4)
13. LOX fill and drain
14. System tunnel (4)
15. Outer LOX tank (4)
16. Central LOX tank
17. Slosh baffle
18. Antenna panel
19. Command antenna (2)
20. Line for gaseous oxygen
21. Instrument section
22. Collar

(SA-201)
Second Stage
S-IVB and IU

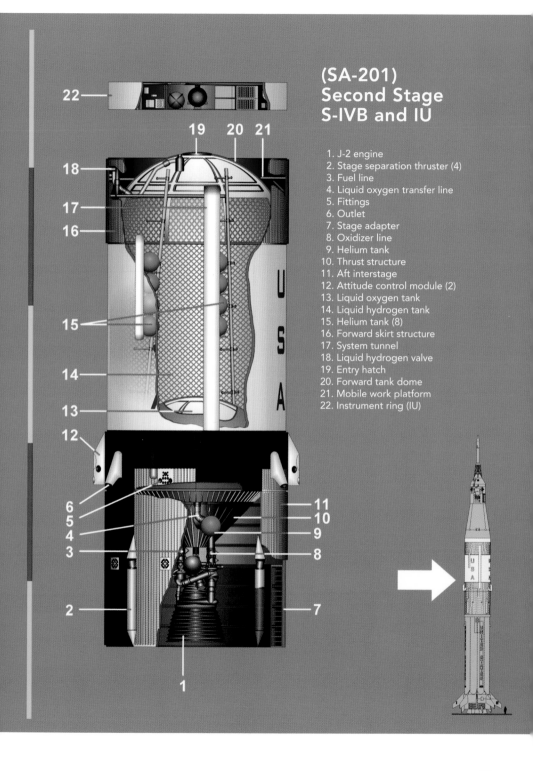

1. J-2 engine
2. Stage separation thruster (4)
3. Fuel line
4. Liquid oxygen transfer line
5. Fittings
6. Outlet
7. Stage adapter
8. Oxidizer line
9. Helium tank
10. Thrust structure
11. Aft interstage
12. Attitude control module (2)
13. Liquid oxygen tank
14. Liquid hydrogen tank
15. Helium tank (8)
16. Forward skirt structure
17. System tunnel
18. Liquid hydrogen valve
19. Entry hatch
20. Forward tank dome
21. Mobile work platform
22. Instrument ring (IU)

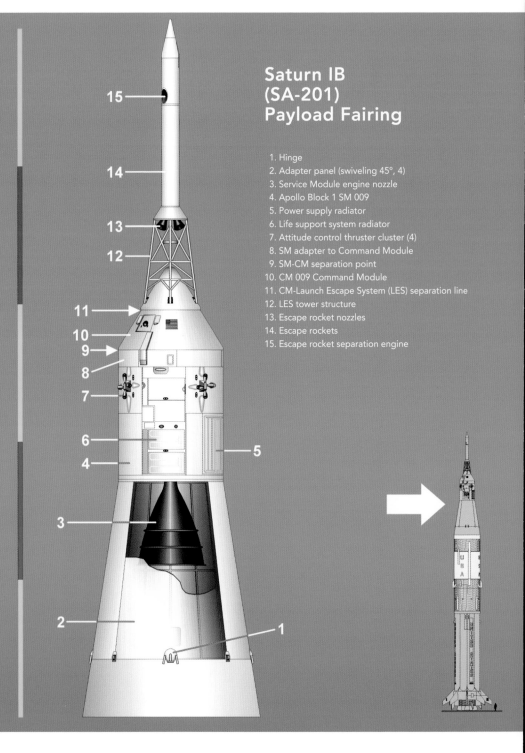

Saturn IB
(SA-201)
Payload Fairing

1. Hinge
2. Adapter panel (swiveling 45°, 4)
3. Service Module engine nozzle
4. Apollo Block 1 SM 009
5. Power supply radiator
6. Life support system radiator
7. Attitude control thruster cluster (4)
8. SM adapter to Command Module
9. SM-CM separation point
10. CM 009 Command Module
11. CM-Launch Escape System (LES) separation line
12. LES tower structure
13. Escape rocket nozzles
14. Escape rockets
15. Escape rocket separation engine

Saturn V SA 503 (Apollo 8)	
Country of Origin	USA
Manufacturer	NASA MSFC
Length	362.86 feet
Diameter over Fins	59 feet
Total Weight	3,230 tons
Total Thrust	7,751,412 pounds
Payload in LEO/ESC	286,600/97,000 pounds
First Stage: Boeing S-IC	
Power Plant	5 x Rocketdyne F-1
Thrust at Sea Level	5 x 1,550,282 pounds
Specific Impulse Sea Level	265 seconds
Specific Impulse Vacuum	304 seconds
Burn Duration	150 seconds
Fuel/Oxidizer	Kerosene/Oxygen
Weight Empty/Fueled	297,624/617,294 pounds
Length/Diameter	138/33 feet
Second Stage: North American S-II	
Power Plant	5 x Rocketdyne J-2
Thrust (total)	231,553 pounds
Specific Impulse	421 seconds
Burn Duration	390 seconds
Fuel/Oxidizer	Hydrogen/Oxygen
Weight Empty/Fueled	79,366/1,058,218 pounds
Length/Diameter	81.5/33 feet
Third Stage: Douglas S-IVB	
Power Plant	1 x Rocketdyne J-2
Thrust	231,553 pounds
Specific Impulse	421 seconds
Burn Duration	475 seconds
Fuel/Oxidizer	Hydrogen/Oxygen
Weight Empty/Fueled	29,321/264,555 pounds
Length/Diameter	61.68/21.6 feet
Apollo Service Module	
Power Plant	1 x Aerojet SPS
Thrust	22,031 pounds
Specific Impulse	530 seconds
Fuel/Oxidizer	Aerozine 50/N2O4
Weight Empty/Fueled	13,227/52,029 pounds
Length/Diameter	14.75/12.8 feet
Adapter and Escape Tower (LES)	
Length Adapter and LES	81.7 feet
Diameter	12.8 - 21.4 feet
Weight Adapter and LES	12,920 pounds
Operational History	
First Launch	November 9, 1967
Launches (Total)/Successful	13/13
Last Launch	May 14, 1973

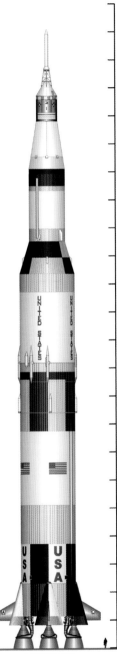

Saturn V (SA-503)
(NL : Apollo 8)

Saturn V
First Stage S-IC

1. F-1 engine (5)
2. Engine skirt (4)
3. Base heat shield
4. Fin (4)
5. Central engine mounting
6. Retrorockets
7. Thrust structure
8. Connection panel
9. Launch clamp (?)
10. Cable tunnel (with fuel pressure line and gaseous oxygen)
11. Oxidizer suction line (5)
12. Fuel tank
13. Slosh baffle
14. Cruciform baffle
15. Helium tanks (4)
16. Liquid oxygen tank
17. Instrument section

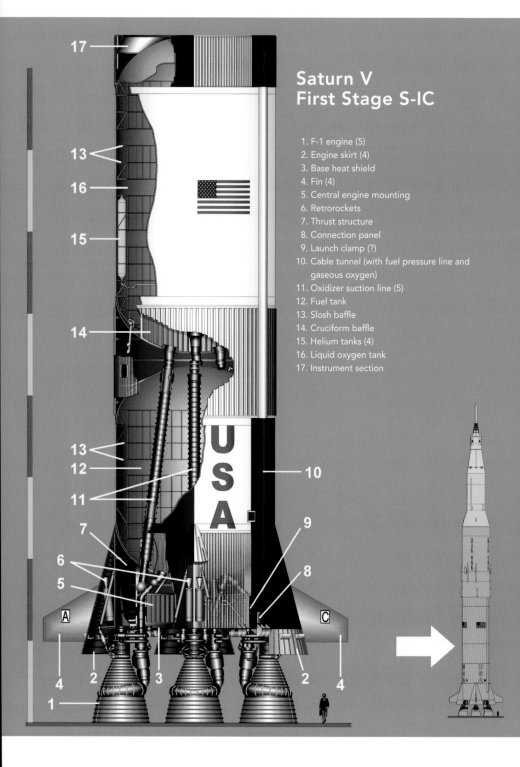

Saturn V
Second Stage S-II

1. J-2 engine (5)
2. Central engine mounting
3. Ullage rocket (8)
4. Thrust structure
5. Slosh baffle
6. Liquid oxygen tank
7. Liquid hydrogen fill and drain
8. Liquid hydrogen suction line
9. Liquid hydrogen feed
10. Liquid oxygen level sensor
11. Liquid oxygen vent line
12. Pressurization mast
13. Tank dome
14. Fuel level sensor
15. Liquid hydrogen tank
16. Work platform
17. Liquid hydrogen vent connect fittings
18. Helium pressure tank (2)
19. Liquid hydrogen diffuser
20. Fuel level sensor

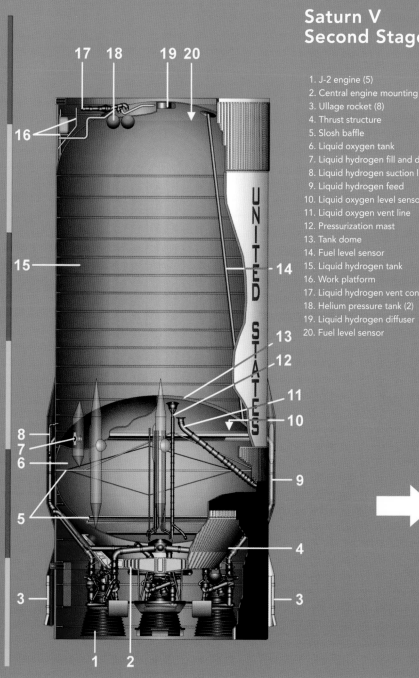

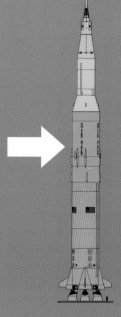

Saturn V
Third Stage
S-IVB and IU

1. J-2 engine
2. Stage adapter
3. Retro motors (4)
4. Helium tanks
5. Aft tank base
6. Auxiliary propulsion system module
7. Liquid oxygen tank
8. Liquid hydrogen tank
9. Forward skirt
10. System tunnel
11. 8 helium tanks (4 per row)
12. Liquid hydrogen valve
13. Entry hatch
14. Forward tank dome
15. Mobile work platform
16. Instrument ring (IU)

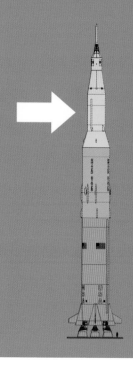

Saturn V
IU and Payload Fairing

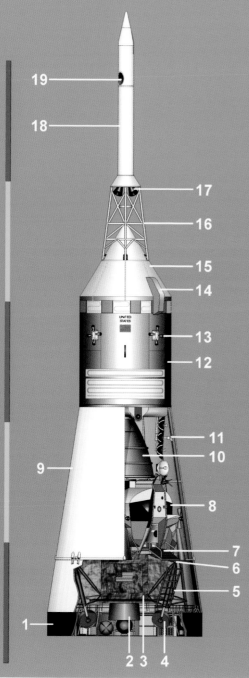

1. Instrumentation Ring (IU)
2. Engine for the Lunar Module descent stage
3. Lunar Module descent stage
4. Landing leg with disc
5. Ladder
6. Lunar Module third stage retaining pins
7. Exit platform
8. Lunar Module ascent stage
9. Jettisonable fairing panels (4) around lander
10. Service Module rocket nozzle
11. High gain antenna (HGA)
12. Supply section (SM)
13. Attitude control thruster cluster (4)
14. Fairing over SM-CM connections
15. Boost protective cover
16. Launch Escape System (LES) tower structure
17. Escape rocket nozzles
18. Escape rockets
19. Escape rockets separation motor

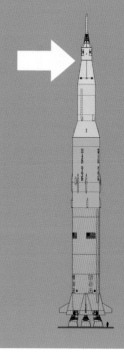

THE 32 MISSIONS
OF THE APOLLO PROGRAM

SATURN I BLOCK 1 – SA 1

The flight of the Saturn A 1 was the first mission by a booster of the Saturn family and the first launch of a Saturn I Block 1.

The second stage and the payload nose cone were functionless dummies, each of which contained a tank holding about 1,412 cubic feet of water to represent the weight proportions of a complete booster. The rocket's first stage was only filled to 83% for this mission.

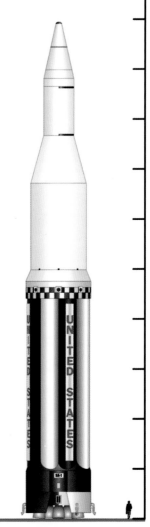

Mission Data and Flight Information SA 1	
Mission Designation	SA 1
Payload	Ballast (water)
Launch Site	LC-34
Ballast Weight	187,400 pounds
Arrival of S-I at the Cape	August 15, 1961
Arrival of S-IV Dummy at the Cape	August 15, 1961
Instrument Unit at the Cape	August 15, 1961
Stacking S-I at Pad 34	August 20, 1961
Stacking S-IV Dummy	August 21, 1961
Stacking Jupiter Nosecone	August 21, 1961
S-I Ignition Command	October 27, 1961, 15:06:01
Range Zero/Launch	15:06:04/000:00:00 MET
S-I Cutoff	000:01:55 MET
Mission Duration	2 min. 40 sec.
Crashed into Atlantic after	14 minutes
Flight Path	Suborbital
Apogee	85 miles
Perigee	200 miles

Information in this and all following mission tables: all mission times in this and the following tables are given in MET, or Mission Elapsed Time. This is the time elapsed after the start of the mission. MET remains the standard expression of time in NASA documents to the present day. It is calculated in hours, minutes and seconds after the time T Zero or Range Zero (000:00:00 MET) and is independent of time zones. For example, termination of burn of the S-I stage took place at 000:01:55 MET, one minute and 55 seconds after the Range Zero command. This command took place exactly at the moment the rocket restraining clamps on the pad released and three minutes after the start of ignition of the first stage engines.

All other times in the tables are given in UTC, or Coordinated Universal Time, equivalent to Greenwich Mean Time. Times in the accompanying text are as a rule given in Cape Canaveral local time.

Saturn I / Block 1
(SA-1)

The Saturn I was a quantum leap in US booster rocket technology. It was three times as tall as the Jupiter, it used six times more fuel, and produced ten times the thrust of the Jupiter. The first stage of the Saturn I was the first space vehicle which, on account of its size, had to be transported by barge to Cape Canaveral. The first stage and the two upper stage mockups reached the base on board the *Compromise*. The trip was difficult. The ship ran aground four times due to inadequate marine charts. On the trip back there was another problem, when the ship struck a bridge.

At the time of the first flight the Saturn concept had not been completely defined. Saturn SA 1 was still conceived as a three-stage vehicle, although the upper stage dummies were nothing more than aerodynamic cases with water tanks inside.

On the day of the launch the curious onlookers waited several miles from Launch Pad 34 for the inevitable. Never before had the maiden flight of a completely new booster rocket gone without incident, and the Saturn I was considerably more complex than any booster before it. At the time the mission was given just a thirty percent chance of proceeding nominally, and that was an optimistic view. Surprisingly the flight came off without a hitch. Prior

The first Saturn I Block 1 with the designation SA 1, built by the Marshall Space Flight Center, leaves Pad 34 at Cape Canaveral Air Force Base, on October 27, 1961.

to launch there were only a few weather-related delays, which totaled one hour.

The only minor problem was burn termination took place 1.6 seconds too early compared to pre-launch calculations. This was traced to sloshing of the remaining fuel in the stage toward the end of the burn phase. The sloshing prevented all of the fuel in the tank from being used.

Here the completed Saturn I Block 1 for the SA 1 mission is at the Fabrication and Assembly Engineering Division at Marshall Space Flight Center. The photo was taken in early February 1961.

SATURN I BLOCK 1 – SA 2

The flight of the Saturn A 2 was the second mission by a booster of the Saturn family and the second launch of a Saturn I Block 1.

The mission was essentially a repeat of the first flight. Once again the second stage and the payload nose cone were dummies containing tanks with a combined total of 3,037 cubic feet of water. The first stage fuel tanks were again only filled to 83% of capacity.

The engineers at the Marshall Space Flight Center had introduced a significant change after the flight of SA 1. They installed additional slosh baffles in the SA 2's kerosene tank to prevent the fuel from sloshing back and forth toward the end of the burn phase. This had caused a premature burn termination on SA 1. The countdown on April 25 went smoothly. There was just one interruption, when a ship entered the flight safety zone sixty miles off the coast and had to be escorted out.

Launch of SA 2 on April 25, 1962

After the eight H-1 engines shut down at an altitude of thirty-five miles, at a flight time of 1 minute and 55 seconds, and a velocity of 3,728 miles per hour, the vehicle continued to rise in unpowered flight to an altitude of 65.5 miles. There the rockets, as planned as

Mission Data and Flight Information SA 2	
Mission Designation	SA 2
Payload	Ballast (water)
Launch Site	LC-34
Ballast Weight	209,439 pounds
Arrival of S-I at the Cape	February 27, 1962
Arrival of S-IV Dummy at the Cape	February 27, 1962
Instrument Unit at the Cape	February 27, 1962
Stacking S-I at Pad 34	February 28, 1962
Stacking S-IV Dummy	March 1, 1962
Stacking Jupiter Nosecone	March 1, 1962
S-I Ignition Command	April 25, 1962, 14:00:31
Range Zero/Launch	14:00:34/000:00:00 MET
S-I Cutoff	000:01:55 MET
Mission Duration	2 min. 40 sec.
Crashed into Atlantic after	13 minutes
Flight Path	Suborbital
Apogee	99 miles
Perigee	214 miles

part of Operation Highwater, exploded and the entire water load was released explosively. The experiment was used to explore the ionosphere, examine the creation of noctilucent clouds and the behavior of ice in space. The resulting cloud was tracked to the apogee of the flight path, at an altitude of approximately ninety-nine miles.

The flight was assessed as a complete success.

A Saturn I S-I under production in the Fabrication and Engineering Laboratory of the Marshall Space Flight Center in Huntsville. The two hub assemblies at opposite ends of the stage have already been attached to the central liquid oxygen tank which was 105 inches in diameter. The four oxygen tanks, with a diameter of 70 inches, have also already been installed. One of the kerosene tanks is just being moved into installation position. Also clearly visible are the forty-eight fiberglass pressure tanks for the gaseous nitrogen, with which the fuel tanks were pressurized, on the right hub assembly (which formed the upper end of the stage).

SA 2 is seen here on Launch Pad 34, at Cape Canaveral Air Force Base, in April 1962.

SATURN I BLOCK 1 – SA 3

The flight of the Saturn A 3 was the third mission of the Apollo program, the third by a booster of the Saturn family, and the third launch of a Saturn I Block 1.

This mission was the first in which the first stage's fuel tanks were completely filled. Once again the main purpose of the flight was to test the H-1 engines in flight, this time with a somewhat longer burn duration. Also tested were new telemetry equipment and a new stabilization platform. Once again the second stage and payload nose cone were dummies, each with a tank, holding a combined total of 3,083 cubic feet of water. The separation engines for separation of the first and second stages were tested, although no active second stage was carried.

The rocket was delivered to the Cape by the barge Promise on September 19. Assembly of the rocket at the launch pad began on September 21. The water ballast was loaded on October 31, and fueling began on November 14, with the RP-1 being loaded first. By the time of the launch the Cuban Crisis was in full swing, and for this reason Kurt Debus, director of NASA launch operations at Cape Canaveral, asked his boss, Wernher von Braun, not to

Saturn SA 3 launches on November 16, 1962.

Mission Data and Flight Information SA 3	
Mission Designation	SA 3
Payload	Ballast (water)
Launch Site	LC-34
Ballast Weight	192,463 pounds
Arrival of S-I at the Cape	September 19, 1962
Arrival of S-IV Dummy at the Cape	September 19, 1962
Instrument Unit at the Cape	September 19, 1962
Stacking S-I at Pad 34	September 21, 1962
Stacking S-IV Dummy	September 24, 1962
Stacking Jupiter Nosecone	September 24, 1962
S-I Ignition Command	November 16, 1962, 17:49:51
Range Zero/Launch	14:00:34/000:00:00 MET
S-I Cutoff Inner Engines	000:02:22 MET
S-I Cutoff Outer Engines	000:02:29 MET
Mission Duration	4 min. 52 sec.
Crashed into Atlantic after	16 minutes
Flight Path	Suborbital
Apogee	104 miles
Perigee	267 miles

invite any official observers, so that the Communist nations would not misinterpret this as a demonstration of power.

The launch followed a forty-five-minute delay caused by a technical defect in a ground system. The four inner H-1 engines were shut down 2 minutes and 22 seconds after leaving the launch pad at an altitude of thirty-eight miles. The four outer engines were kept running for a "burn to depletion," or until one of the two fuel components was expended. This took place 2 minutes and 29 seconds after liftoff at an altitude of forty-four miles and a velocity of 4,040 miles per hour, when the oxidizer became the first component to be used up. Two minutes and 34 seconds after liftoff the separation engines fired for 2.1 seconds. Four minutes and 52 seconds after liftoff, at an altitude of 104 miles and a distance of 131 miles from the launch site, the second stage and payload section dummies were blown up and their contents, water, were released for Project Highwater. The first stage remained intact and crashed into the Atlantic 267 miles from the launch point.

This photo was taken during SA 1 launch preparations, but the procedure for SA 3 was exactly the same. Here the rocket is being anchored in the hold-down arms.

As soon as the vehicle was fixed stably to the launching base, installation of the upper stage dummies with the water tanks began. Visible here on the base of the launching base are three of the four hold-down arms (pale grey, at ninety degree intervals). They held the rocket down on the launching base during the three seconds of thrust buildup and then released it by pushing away from the rocket. This photo shows SA 4, however the procedure was exactly the same for SA 3.

SATURN I BLOCK 1 – SA 4

Saturn SA 4 was the fourth booster of the Saturn family and the fourth and last flight by a Saturn I Block 1.

The suborbital mission was also the last flight by a Saturn rocket in which only the first stage was active. It was the mission with the shortest check-out time (just fifty-four days) to date and the longest countdown delay (120 minutes), caused by technical problems. Essentially the mission was a repeat of the third mission. The dummy second stage looked exactly like a Block 2 second stage, including all ventilation shafts, fairings, and even the dummy camera pod.

The focal point of this test, however, was the behavior of the booster in the event of an engine failure. The plan called for engine number five to be shut down after 100 seconds of flight. The rocket was then supposed to independently compensate for the loss of an engine with a longer burn by the other engines and divert the necessary fuel.

As envisaged, engine number five shut down after 100 seconds of flying time. It had been suspected that the shut-down engine might disintegrate because of the heat production from the engines

**Saturn I / Block 1
(SA-4)**

Mission Data and Flight Information SA 4	
Mission Designation	SA 4
Payload	Ballast (water)
Launch Site	LC-34
Ballast Weight	99,200 pounds
Arrival of S-I at the Cape	February 2, 1963
Arrival of S-IV Dummy at the Cape	February 2, 1963
Instrument Unit at the Cape	February 2, 1963
Stacking S-I at Pad 34	February 4, 1963
Stacking S-IV Dummy	February 5, 1963
Stacking Jupiter Nosecone	February 5, 1963
S-I Ignition Command	March 28, 1963, 20:26:52
Range Zero/Launch	20:26:55/000:00:00 MET
S-I Cutoff Engine 5	000:01:40 MET
S-I Cutoff Inner Engines	000:02:21 MET
S-I Cutoff Outer Engines	000:02:27 MET
Mission Duration	15 min.
Flight Path	Suborbital
Apogee	80 miles
Perigee	250 miles

This photo shows the main production building of NASA's Manned Space Flight Center, in Huntsville, Alabama. Three Saturn I first stages can be seen in various stages of completion. From left to right are: SA 6, SA 7, and SA 4.

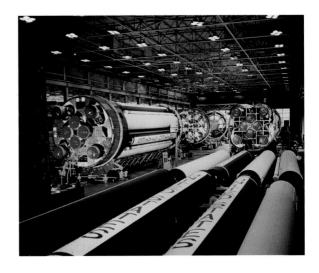

that were still working, after the cooling circuit was interrupted. This was not the case, however. The mission was an important test in proving the operational feasibility of engine clustering. After burn termination the stage separation motors were also ignited, as if an active second stage was being carried. In fact, however, there was no separation from the first stage.

The first stages of all Saturn variants, the S-I, the S-IB and the S-IC used a combination of RP-1 (Refined Petroleum, sometimes also called Rocket Propellant 1) and liquid oxygen as oxidizer. RP-1 was a further refined form of JP-4 (jet fuel), which was used in aviation. In 1958, when the decision was made to develop the basic stage of the Saturn I, there were already good experiences and a multi-year operational history with this fuel.

Preparations are made to upright SA 4 on the launch pad.

SATURN I BLOCK 2 – SA 5

Saturn SA 5 was the fifth booster of the Saturn family and the first flight of a Saturn I Block 2. Launch Complex 37A was also inaugurated with this launch.

For the first time a Saturn rocket was launched with the S-IV second stage, which was powered by six RL-10A-3 engines. The first stage were larger, the engines more powerful. The first stage was equipped with eight fins to improve stability during the atmospheric flight phase.

After many delays, mainly technical in nature, the launch was scheduled for January 27, 1964; but then at the last minute there was a problem with the fueling system and the mission had to be postponed for two days. On January 29, there were further minor problems, but at 1125 Eastern Time, the moment arrived. The ascent into orbit was perfect. This mission was documented like none before it. The flight of the booster rocket was observed by nineteen ground-based cameras, four with large Baker-Nunn lenses, which precisely recorded every movement. The stage separation was filmed by a total of eight cameras on board the Saturn. In the age of vacuum tubes this was a sensation. Their primary purpose was

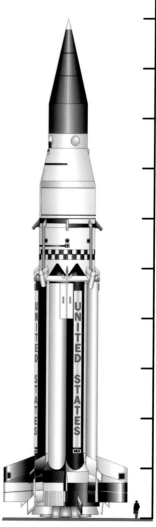

Saturn I / Block 2 (SA-5)

Mission Data and Flight Information SA 5	
Mission Designation	SA 5
Payload	Ballast
Launch Site	LC-37B
Ballast Weight	23,369 pounds
Arrival of S-I at the Cape	August 21, 1963
Arrival of S-IV Dummy at the Cape	September 21, 1963
Instrument Unit at the Cape	August 21, 1963
Stacking S-I at Pad 37B	August 25, 1963
Stacking S-IV	October 11, 1963
Stacking Instrument Unit	October 14, 1963
Stacking Payload Dummy	October 22, 1963
S-I Ignition Command	January 29, 1964, 16:24:58
Range Zero/Launch	16:25:01/000:00:00 MET
S-I Cutoff Central Engines	000:02:21 MET
S-I Cutoff Outer Engines	000:02:27 MET
S-I/S-IV Separation	000:02:29 MET
S-IV Ignition Command	000:02:30 MET
S-IV Ullage Case Jettisoned	000:02:41 MET
S-IV End of Burn	000:10:25 MET
Mission Duration	15 min.
Crash back to Earth	April 30, 1966

to closely observe one event: the first stage separation by a Saturn rocket and the first ignition of a second stage in the program. After separation of the stage they were jettisoned and came down by parachute. They were picked up in the Atlantic, 500 miles from the launch site. Stage separation was perfect. The S-IV upper stage engines continued running for eight more minutes, placing it and the payload dummy into orbit. With a weight of just under 37,500 pounds, at that time it was the most massive object ever placed into orbit.

The test engineers had equipped the rocket with hundreds of sensors, which took data readings at 1,200 different places. The results were transmitted to the Cape and to seven other receiving stations along the flight path. Separation of the S-I from the S-IV began 141 seconds after the rocket left the launch pad. At that time one of the first stage's inner engines was shut down. The other seven H-1 engines ran for another six seconds, before they were also shut down. At 0.2 seconds later small solid-fuel motors on the upper stage were ignited in order to maintain slight acceleration during the brief free-flight phase between the shutdown of the first stage and the ignition of the second stage, so that the fuel in the upper stage remained at the base of the tank. Then additional solid-fuel retro rockets ignited on the first stage to slow it. At the same time an explosive cord detonated, shearing off the bolts holding the two stages together. Immediately afterward the second stage's six RL-10 engines ignited. The two units separated from each other quickly. The first stage followed its ballistic flight path and reached an altitude of ninety miles. The camera pods were also separated at this time.

The launch of SA 5 was broadcast live on television all over America. The cameras followed the rocket until stage separation at a height of thirty-five miles. American President Johnson also watched the launch on television and subsequently congratulated the team in the blockhouse at Pad 37B by telephone. George Mueller, the director of NASA's manned spaceflight program, described the launch as "the first step to the moon."

The launch of SA 5 on the first orbital mission of the Apollo program, on January 29, 1964. It was the first mission from Launch Complex 37B.

Artist's drawing of the S-IV upper stage with Jupiter payload nosecone entering orbit.

SATURN I BLOCK 2 – SA 6

Saturn SA 6 was the sixth booster of the Saturn family and the second flight by a Saturn I Block 2. It was the second orbital mission of the Apollo program and the first in which a boilerplate Apollo was placed into orbit.

The assembly of the stages, instrument unit, and payload of SA 6 began at Launch Pad 37B at Cape Canaveral Air Base on Thursday, February 20, 1964.

The first five Saturn I rockets were still equipped with the payload fairings of the Jupiter medium-range missile. The main purpose of this mission therefore was verification of the launch loads on an Apollo Command and Service Module including escape tower (LES). Apollo Boilerplate No. 13 was fitted with 116 sensors, which transmitted their data by telemetry to the ground stations.

Whereas launch preparations for SA 5 had lasted more than 150 days all in all, preparations for SA 6 took place in just ninety-one days. The launch followed on May 28, 1964, twice having been called off because of technical problems. The ascent by the booster was nominal until 117 seconds into the flight. Then, engine number eight unexpectedly shut down. The rocket compensated by extending

Saturn I Block 2 AS 101 waits at Complex 37B for its mission to begin.

the burn time of the seven remaining engines by three seconds, as provided for in the design. Ten seconds after stage separation the launch escape system was jettisoned as planned, as were the eight film cameras which had observed the separation process. Burn termination followed 625 seconds after leaving the launch pad. The stage and boilerplate remained together physically and transmitted data to the ground stations for about six seconds, until the batteries were discharged. The vehicle circled the earth a total of fifty-four times before it burned up over the Phoenix Islands in the Pacific. Recovery of the capsule was neither planned nor would it have been possible, for the Boilerplate had no heat shield. Contact was maintained with it for almost six hours, however.

Mission Data and Flight Information SA 6	
Mission Designation	AS 101
Payload	Apollo Boilerplate 13 (BP 13)
Launch Site	LC-37B
Combined Payload Weight	36,817 pounds
Arrival of S-I at the Cape	February 19, 1964
Arrival of S-IV at the Cape	February 22, 1964
Instrument Unit at the Cape	February 19, 1964
Stacking S-I at Pad 37B	February 20, 1964
Stacking S-IV	March 19, 1964
Stacking Instrument Unit	March 23, 1964
Stacking with Payload	April 2, 1964
S-I Ignition Command	May 28, 1964, 17:06:57
Range Zero/Launch	17:07:00/000:00:00 MET
S-I Cutoff Engine 8	000:01:57 MET
S-I Cutoff Other Engines	000:02:30 MET
S-I/S-IV Separation	000:02:32 MET
S-IV Ignition Command	000:02:33 MET
S-IV Ullage Case Jettisoned	000:02:41 MET
Escape Tower Jettisoned	000:02:41 MET
S-IV End of Burn	000:10:25 MET

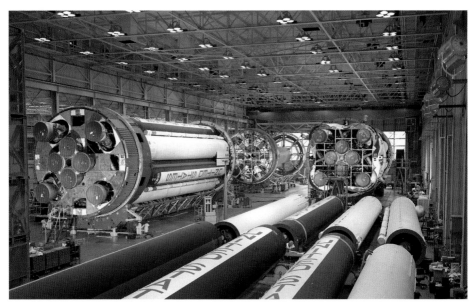

In this photo can be seen the Saturn I S-I stages for SA 4, SA 6 (AS 101), and SA 7 (AS 102) under construction in Building 4705 at the Fabrication and Assembly Engineering Division of the Marshall Space Flight Center. The photo was taken on January 13, 1963.

The cause of the premature shutdown of engine number eight was quickly found. A tooth had broken off in one of the drives in the turbopumps. This did not result in any delays in the program, however, for the weak spot was already known and a modified design was already in preparation for the next launches. It was the only problem encountered with the H-1 engine during the Saturn I missions.

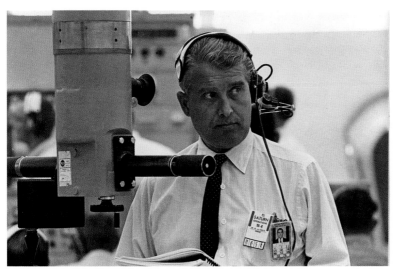

Wernher von Braun, director of the Marshall Space Flight Center, at the periscope in the control bunker at Launch Complex 37B, on May 28, 1964.

SATURN I BLOCK 2 – SA 7

S aturn SA 7 was the seventh booster of the Saturn family and the third flight by a Saturn I Block 2. It was the third orbital mission of the Apollo program and the second in which a boilerplate Apollo was placed into orbit.

The assembly of the main components of SA 7 began at Launch Pad 37B at Cape Canaveral Air Base on Thursday, June 9, 1964.

The SA 7 mission was essentially a repeat of SA 6 with additional sensors on board. These were largely associated with the function of the Apollo Boilerplate. AS 102 was the first Saturn to carry a programmable computer. All of the previous missions had carried a black box which was preprogrammed. Computer entries were now possible during countdown and, theoretically, even during the flight. On this mission the Apollo also included a partially active escape tower, which was jettisoned during the ascent phase.

A small crack in an H-1 first stage engine was discovered in early July. NASA subsequently decided to replace all eight engines and send them back to the manufacturer for inspection. This delayed

Mission Data and Flight Information SA 7	
Mission Designation	AS 102
Payload	Apollo Boilerplate 15 (BP 15)
Launch Site	LC-37B
Combined Payload Weight	36,817 pounds
Arrival of S-I at the Cape	June 7, 1964
Arrival of S-IV at the Cape	June 12, 1964
Instrument Unit at the Cape	June 7, 1964
Stacking S-I at Pad 37B	June 9, 1964
Stacking S-IV	June 19, 1964
Stacking Instrument Unit	June 22, 1964
Stacking with Payload	June 26, 1964
Start of Countdown	September 18, 1964
S-I Ignition Command	September 18, 1964, 16:22:40
Range Zero/Launch	16:22:43/000:00:00 MET
S-I Cutoff Inner Engines	000:02:21 MET
S-I Cutoff Other Engines	000:02:27MET
S-I/S-IV Separation	000:02:28 MET
S-IV Ignition Command	000:02:29 MET
S-IV Ullage Case Jettisoned	000:02:39 MET
Escape Tower Jettisoned	000:02:40 MET
S-IV End of Burn	000:10:21 MET

Launch of SA 7 with AS 102, on September 18, 1964.

At the Marshall Space Flight Center, the Saturn I S-IV stage for mission SA 7 is prepared for transport to Cape Canaveral. Note the six RL-10-A3 engines mounted on the thrust structure.

the launch by two weeks. Then Hurricanes Cleo and Dora led to further delays. The launch achieved all of the envisaged objectives. The Boilerplate transmitted telemetry for seven and a half hours. During the fifty-ninth orbit the rocket reentered the earth's atmosphere over the Indian Ocean and burned up. Here again, recovery was neither possible nor planned.

The mission's only anomaly was the loss of the eight film capsules, which could not be found. They landed outside the planned area and Hurricane Gladys made recovery impossible. Two of the pods were washed up on a beach two months later, however. While covered in barnacles, the film inside was undamaged.

The mission SA 7 S-I stage being set up at Launch Complex 37B.

SATURN I BLOCK 2 – SA 9

S aturn SA 9 was the eighth booster of the Saturn family and the fourth flight by a Saturn I Block 2. It was the fourth orbital mission of the Apollo program, the third in which a boilerplate Apollo was placed into orbit, and the first with a Pegasus payload. It was the first flight of a fully operational Saturn I.

The assembly of the main components of SA 9 began at Launch Pad 37B at Cape Canaveral Air Base on Thursday, November 3, 1964.

The payload consisted of Apollo Boilerplate spacecraft BP-16, whose dummy Service Module served as a protective cover for the Pegasus A satellite, the first in a series of three micro-meteorite detectors used for the Apollo program. Pegasus A alone weighed 3,980 pounds.

The Saturn first stage Number 9 was the last unit built by the Marshall Space Flight Center. The first stages of units SA 8, SA 10, and all following Saturn IB rockets were built by Chrysler at the

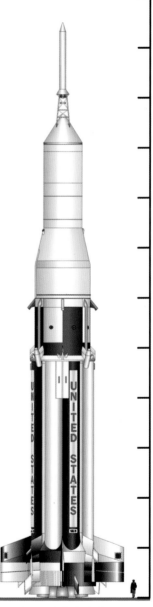

**Saturn I / Block 2
(SA-9)**

Mission Data and Flight Information AS 103	
Mission Designation	AS 103
Payload 1	Apollo Boilerplate 16 (BP 16)
Payload 2	Pegasus A
Launch Site	LC-37B
Combined Payload Weight	33,896 pounds
Arrival of S-I at the Cape	October 30, 1964
Arrival of S-IV at the Cape	October 23, 1964
Instrument Unit at the Cape	October 30, 1964
Stacking S-I at Pad 37B	November 3, 1964
Stacking S-IV	November 19, 1964
Stacking Instrument Unit	November 19, 1964
Stacking with Payload	January 14, 1965
Start of Countdown	February 12, 1965
S-I Ignition Command	February 16, 1965, 14:37:00
Range Zero/Launch	14:37:03/000:00:00 MET
S-I Cutoff Inner Engines	000:02:20 MET
S-I Cutoff Other Engines	000:02:26 MET
S-I/S-IV Separation	000:02:26 MET
S-IV Ignition Command	000:02:28 MET
S-IV Ullage Case Jettisoned	000:02:38 MET
Escape Tower Jettisoned	000:02:38 MET
S-IV End of Burn	000:10:22 MET
Separation of BP 16 from S-IV	000:13:24 MET

This photo was taken at the beginning of November 1964. Here, in a hangar at Cape Canaveral, the Saturn I S-IV stage for mission SA 9 is weighed and balanced in preparation for assembly at Launch Complex 37B.

NASA facilities in Michoud, Louisiana. After Chrysler fell behind in its deliveries, however, SA 9, the last NASA-produced unit, was launched before SA 8. The Pegasus payload was housed in the empty hull of the Boilerplate 16 Service Module. It remained in orbit attached to the booster's second stage. This large sensor was used to investigate the frequency of meteorites in low-earth orbit.

Launch took place at 0937 local time. Previously there had been an interruption of one hour and seven minutes in the countdown because of problems with a computer at the Eastern Test Range. The launch was completely normal and the combination of boilerplate, Pegasus, and upper stage reached orbit ten minutes and thirty seconds after liftoff. The boilerplate capsule and the capsule fairings separated from the S-IV stage thirteen minutes and twenty-four seconds after launch. The two micrometeorite detectors deployed one minute later and the first hit by a micrometeorite was recorded during the fourth orbit of the earth.

The meteorite sensor continued to operate for 1,188 days. On August 29, 1968, it was shut down. The spacecraft remained in orbit until July 10, 1985, when it entered the earth's atmosphere and burned up.

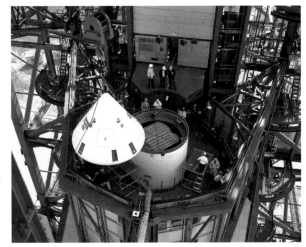

At Launch Complex 37B it is quite clear that the Apollo AS 103 Service Module is merely a shell. Folded inside is the Pegasus A particle detector. As this picture was being taken, Apollo Boilerplate 16 was being fitted.

SATURN I BLOCK 2 – SA 8

S aturn SA 8 was the ninth booster of the Saturn family and the fifth flight by a Saturn I Block 2. It was the fifth orbital mission of the Apollo program, the fourth in which a boilerplate Apollo was placed into orbit, and the second with a Pegasus payload. It was the second flight of a fully operational Saturn I.

Assembly of the main components of SA 8 began at Launch Pad 37B at Cape Canaveral Air Force Base on March 2, 1965.

The payload consisted of Apollo Boilerplate spacecraft BP-26, whose dummy Service Module was to serve as a protective cover for the Pegasus B satellite, the second in a series of three micrometeorite detectors deployed in support of the Apollo program. The Pegasus satellite alone weighed 3,086 pounds.

Saturn first stage number 9 was the last unit built by the Marshall Space Flight Center. The first stage units for the SA 8, SA 10, and all following Saturn IB rockets were built by Chrysler in Michoud, Louisiana. After Chrysler fell behind in its deliveries, SA 9, the last NASA-built unit, was launched before SA 8. The Pegasus

A time exposure of the launch of Saturn SA 8. Significant parts of the infrastructure for the coming Saturn V flights are present in this photo. On the right is the Vehicle Assembly Building, which had just been completed, and to its left three launcher umbilical towers.

Mission Data and Flight Information AS 104	
Mission Designation	AS 104
Payload 1	Apollo Boilerplate 26 (BP 26)
Payload 2	Pegasus B
Launch Site	LC-37B
Combined Payload Weight	33,951 pounds
Arrival of S-I at the Cape	February 28, 1965
Arrival of S-IV at the Cape	February 26, 1965
Instrument Unit at the Cape	March 8, 1965
Stacking S-I at Pad 37B	March 2, 1965
Stacking S-IV	March 17, 1965
Stacking Instrument Unit	March 17, 1965
Stacking with Payload	April 28, 1965
Start of Countdown	May 20, 1965
S-I Ignition Command	May 25, 1965, 07:34:58
Range Zero/Launch	07:35:01/000:00:00 MET
S-I Cutoff Inner Engines	000:02:21 MET
S-I Cutoff Other Engines	000:02:28 MET
S-I/S-IV Separation	000:02:29 MET
S-IV Ignition Command	000:02:31 MET
S-IV Ullage Case Jettisoned	000:02:41 MET
Escape Tower Jettisoned	000:02:41 MET
S-IV End of Burn	000:10:24 MET
Separation of BP 26 from S-IV	000:13:26 MET

Scene from the launch bunker at Complex 37B. In the center of the photo (pointing) is Kurt Debus, then NASA's Director of Launch Operations (later director of the Kennedy Space Center). To the right of him is Wernher von Braun.

The launch of SA 8 at 0235 Eastern Time on May 25, 1965, was one of just two night launches of a Saturn rocket.

payload was accommodated in the empty hull of the Boilerplate 26 Service Module. It remained in orbit attached to the booster's second stage. This large sensor examined the frequency of meteorites in low-earth orbit.

The S-IV stage with Pegasus B achieved an orbit with a lowest orbital point of 317.5 miles and a highest point of 460 miles. It reentered earth's atmosphere on July 8, 1989.

The Pegasus satellites played an important role in the Apollo program, but also in the development of the Saturn rockets. It proved that micrometeorites were much less common in low-earth orbit than previously thought. As a result, special measures against this threat were for the most part dropped. Consideration had originally been given to fitting the S-IVB with a meteorite shield. This extremely weight-intensive measure was thus not needed.

SATURN I BLOCK 2 - SA 10

The launch of SA 10 from Launch Complex 37B, on July 30, 1965.

Saturn SA 10 was the tenth booster of the Saturn family and the sixth and last flight by a Saturn I Block 2. It was the sixth orbital mission of the Apollo program, the fifth in which a boilerplate Apollo was placed into orbit, and the third and last with a Pegasus payload. It was the third flight of a fully operational Saturn I.

Assembly of the main components of SA 10 began at Launch Pad 37B at Cape Canaveral Air Force Base on June 2, 1965.

The payload consisted of the Apollo Boilerplate spacecraft BP-9A, whose dummy Service Module served as protective cover for the Pegasus C satellite, the third in a series of micrometeorite detectors used in support of the Apollo program.

The booster rocket was largely similar to the SA 8 and SA 9 boosters. Its launch weight was 564 tons.

The payload weight given in the table includes the spacecraft, the Pegasus C satellite, the instrument ring, and the S-IV upper stage.

Mission Data and Flight Information SA 10	
Mission Designation	AS 105
Payload 1	Apollo Boilerplate 9A (BP 9A)
Payload 2	Pegasus C
Launch Site	LC-37B
Combined Payload Weight	34,436 pounds
Arrival of S-I at the Cape	May 31, 1965
Arrival of S-IV at the Cape	May 10, 1965
Instrument Unit at the Cape	June 1, 1965
Stacking S-I at Pad 37B	June 2, 1965
Stacking S-IV	June 8, 1965
Stacking Instrument Unit	June 9, 1965
Stacking with Payload	July 6, 1965
Start of Countdown	July 27, 1965
S-I Ignition Command	July 30, 1965, 07:59:57
Range Zero/Launch	08:00:00/000:00:00 MET
S-I Cutoff Inner Engines	000:02:23 MET
S-I Cutoff Other Engines	000:02:30 MET
S-I/S-IV Separation	000:02:30 MET
S-IV Ignition Command	000:02:32 MET
S-IV Ullage Case Jettisoned	000:02:42 MET
Escape Tower Jettisoned	000:02:42 MET
S-IV End of Burn	000:10:42 MET
Separation of BP 26 from S-IV	000:13:32 MET

This photo was taken in November 1964. It shows the Saturn I S-I stages for the SA 8 and SA 10 missions in final assembly at the Michoud Assembly Facility in New Orleans, Louisiana.

The S-IV stage with the Pegasus C achieved an orbit with a perigee of 332 miles and an apogee of 352 miles. It reentered the atmosphere on August 4, 1969.

The Saturn program had created new branches of industry, which slowly took shape during the first flights of the S-IVB. Liquid hydrogen, for example, was made in only tiny laboratory quantities at the end of the 1950s. Each pound cost nine dollars, equivalent to eighty of today's dollars. Thanks to the space program by the mid-1960s daily production had risen to 209 tons, of which 95% was used by the space program. And in the end the price of liquid hydrogen per pound had fallen to sixteen cents per pound.

The Apollo Boilerplate capsule 9A and Pegasus C (in the Service Module shell beneath the capsule) are attached to the Saturn SA 10 at Launch Complex 37B.

SATURN IB – SA 201

S aturn SA 201 was the eleventh booster of the Saturn family and the first flight of a Saturn IB. It was the first launch in which a Block 1 Apollo, CSM 009, was used, and the first in which a production Block 1 Service Module was carried.

The mission was designed as a suborbital flight. On Wednesday, August 18, 1965, SA 201 was initially equipped and assembled with an S-IVB dummy for the Saturn V mockup 500 F and a dummy instrument ring, to check out the newly-installed facilities at Launch Complex 34.

A three-day countdown began on February 20, 1966. As with all maiden flights, delays due to technical problems were expected. As it turned out, the weather at the Cape was the main problem. The countdown had to be stopped twice and then resumed. On February 25, at 1715 local time, the countdown was begun a third time. The next morning at 0900, three seconds before ignition, the booster rocket's computer reported that the pressure in the helium

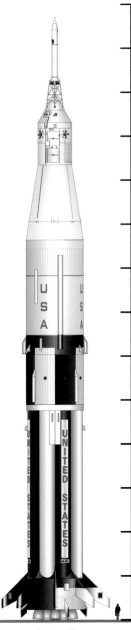

Saturn IB
(SA-201)

Mission Data and Flight Information SA 201	
Mission Designation	Apollo Saturn (SA) 201
Payload 1	Apollo CSM-009
Payload 2	Pegasus C
Launch Site	LC-34
Combined Payload Weight	45,900 pounds
Arrival of S-IB at the Cape	August 14, 1965
Arrival of S-IVB at the Cape	September 19, 1965
Instrument Unit at the Cape	October 20, 1965
Stacking S-IB at Pad 34	August 18, 1965
Stacking S-IVB	October 1, 1965
Stacking Instrument Unit	October 25, 1965
Stacking with CSM-009	December 26, 1965
Start of Countdown	February 21, 1966, 00:00
Start of Final Countdown	February 25, 17:15
S-IC Ignition Command	February 26, 1966, 16:11:58
Range Zero/Launch	16:12:01/000:00:00 MET
S-IB Cutoff Inner Engines	000:02:21 MET
S-IB Cutoff Other Engines	000:02:27 MET
S-IB/S-IVB Separation	000:02:28 MET
S-IVB Ignition Command	000:02:29 MET
S-IVB Ullage Case Jettisoned	000:02:40 MET
Escape Tower Jettisoned	000:02:53 MET
S-IVB End of Burn	000:10:03 MET
Loss of S-IVB Telemetry Signal	000:26:21 MET

A Saturn IB first stage on its way to acceptance firing. The tests took place on a modified S-I test stand at the Marshall Space Flight Center in Huntsville.

supply for the tank pressure regulating system had fallen below the tolerance level. The countdown clock was reset to fifteen minutes before launch and then stopped. Now began discussions between the engineers in Huntsville under Wernher von Braun and those at the Cape under Kurt Debus. After no one at the Cape could say with certainty how serious the problem was, at 1045 the launch was cancelled.

Just minutes later, Wernher von Braun's team from Huntsville called and said that they could continue with the launch it they could return the helium pressure to its original level by feeding from the outside. Von Braun's people suspected that the loss of pressure had been caused by the testing equipment itself and not by a problem with the rocket. Kurt Debus' crew agreed and the countdown was resumed at 1057.

At 1112 the eight first stage engines ignited and took the rocket with the spacecraft to an altitude of thirty-five miles and a speed of 3,600 miles per hour. Then the first stage was separated, the S-IVB upper stage took over subsequent propulsion and delivered the payload to an altitude of 264 miles. There the stage and the spacecraft separated.

Here technicians at the Michoud Assembly Facility (MAF) install the H-1 engines in the S-IB stage of a Saturn IB.

SATURN IB – SA 203

S aturn SA 203 was the twelfth booster of the Saturn family and the second flight of a Saturn IB. It was the first flight of a Saturn IB not carried out with an Apollo Command and Service Module as payload. After delays developed with the Command and Service Module destined for SA 202, the SA 203 mission was moved up.

This mission was quite special in that the rocket carried no payload. The objective was to reach orbit with as much fuel remaining as possible. Event the amount of oxidizer was reduced so that the remaining quantity of hydrogen came as close as possible to the load of an S-IVB in orbit launched by a Saturn V. An arrangement of two cameras and eighty-eight sensors would observe the behavior of the hydrogen in the tanks and measure the temperatures in the fuel lines and engines. This information was important, because a slightly modified version of the S-IVB second stage was also supposed to be used as the third stage of the Saturn V. In this version it had to be reignited several hours after achieving orbit, to deliver the Apollo Command and Service Module and Lunar Module to the moon.

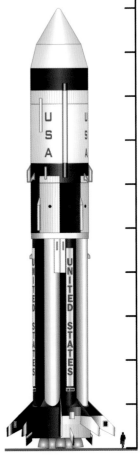

Saturn IB
(SA-203)

Mission Data and Flight Information SA 203	
Mission Designation	Apollo Saturn (SA) 203
Payload	Hydrogen
Launch Site	LC-37B
Arrival of S-IB at the Cape	April 12, 1966
Arrival of S-IVB at the Cape	April 6, 1966
Instrument Unit at the Cape	April 14, 1966
Stacking S-IB at Pad 37B	April 18, 1966
Stacking S-IVB	April 21, 1966
Stacking Instrument Unit	April 21, 1966
Mounting of Nose Cone	April 21, 1966
Start of Final Countdown	July 5, 1966, 01:30
S-IC Ignition Command	July 5, 1966, 14:53:10
Range Zero/Launch	14:53:13/000:00:00 MET
S-IB Cutoff Center Engines	000:01:59 MET
S-IC Cutoff Outer Engines	000:02:03 MET
S-IB/S-IVB Separation Command	000:02:23 MET
S-IVB Ignition Command	000:02:25 MET
S-IVB Ullage Case Jettisoned	000:02:15 MET
Nose Cone Jettisoned	000:02:55 MET
S-IVB End of Burn	000:07:13 MET

liquid for four orbits using the cameras installed in the tanks.

The primary test objectives were carried out during the first two orbits. The liquid hydrogen in the tanks behaved largely as predicted. The next two orbits were used for additional tests to acquire information for the design of future cryogenic stages. These included an experiment in which a pressure valve in the oxygen tank was opened, while at the same time a tank pressure regulating valve was closed in the hydrogen tank above it. It was expected that the partition between them would collapse because of the pressure differential of thirty-nine pounds per square inch. During a 150-second-long interruption in radio contact during the handoff from the Manned Spaceflight Center to the tracking station in Trinidad, however, something happened that could not be reconstructed with certainty. The radar in Trinidad did, however, determine that the rocket had broken into numerous pieces. Telemetry contact could not be reestablished. NASA suspected that a spark in the interior or the impact of a foreign body must have caused the rocket to explode. Despite this the mission was assessed as a complete success.

Launch of SA 203 from Launch Complex 37B.

Compared to other Saturn IB launches, during this launch all the events after range zero took place significantly earlier, which was due to the absence of a payload. The same was true of the countdown itself, which was significantly shorter.

The S-IVB reached orbit with 18,960 pounds of liquid hydrogen remaining. The total weight of the S-IVB stage and fuel was 58,422 pounds. It was the greatest weight so far placed in orbit. The ground stations were able to observe the behavior of the

Wernher von Braun, director of the Marshall Space Flight Center, watches the launch through a periscope in the blockhouse of Launch Complex 37B.

SATURN IB – SA 202

Saturn SA 202 was the thirteenth booster of the Saturn family and the third flight of a Saturn IB. It was the third flight of a Saturn IB with a Block I Apollo. The flight of SA 202 was carried out after the mission of SA 203, because delays had arisen in delivery of the Command and Service Module CSM-011 destined for SA 202.

The mission was designed as a suborbital flight in which expanded performance data was to be gained compared to SA 201.

The majority of the countdown activities were taken care of during the two-part Countdown Demonstration Test on August 2 and 3, and on August 7 and 8. These included two final countdowns, during which the readiness of all systems was verified. All that remained for the twelve-hour actual countdown were purely the launch preparation activities.

Mission SA 202 was considerably more complicated that the mission of SA 201. This time it lasted well over ninety minutes. This time the unmanned Apollo was to reach an altitude of more than 600 miles and carry out two thirds of an earth orbit. The mission

SA 202, the second Saturn IB, launches from Launch Complex 37B at Cape Canaveral, on August 25, 1966. A total of nine Saturn IB flights were carried out, the last as part of the Apollo Soyuz Test Project (ASTP), in July 1975.

Mission Data and Flight Information SA 202	
Mission Designation	Apollo Saturn (SA) 202
Payload	Apollo CSM-011
Launch Site	LC-34
Payload Weight	44,291 pounds
Arrival of S-IB at the Cape	February 7, 1966
Arrival of S-IVB at the Cape	January 29, 1966
Instrument Unit at the Cape	February 21, 1966
Stacking S-IB at Pad 37B	March 4, 1966
Stacking S-IVB	March 10, 1966
Stacking Instrument Unit	March 11, 1966
Start of Final Countdown	August 25, 1966, 03:30
S-IC Ignition Command	August 25, 1966, 17:15:29
Range Zero/Launch	17:15:32/000:00:00 MET
S-IB Cutoff Center Engine	000:02:21 MET
S-IC Cutoff Outer Engines	000:02:24 MET
S-IB/S-IVB Separation Command	000:02:25 MET
S-IVB Ignition Command	000:02:26 MET
S-IVB Ullage Case Jettisoned	000:02:36 MET
Escape Tower Jettison	000:02:50 MET
S-IVB End of Burn	000:09:50 MET
Separation of LM from S-IVB	000:09:59 MET
Loss of S-IVB Telemetry Signal	000:15:41 MET
Verlust S-IVB Telemetriesignal	000:15:41 MET

This photo was taken on June 18, 1965. Technicians of the Marshall Space Flight Center hoist a test version of the S-IVB stage into the dynamic test stand. The entire Saturn IB and its structural integrity were tested in this facility.

included a whole series of difficult tasks. The most important were four burn maneuvers by the Service Module with a total duration of three minutes and twenty seconds. The ignitions functioned perfectly, however the descent maneuver was a little steeper than planned. As a result, the landing took place 235 miles from the target, and it took the aircraft carrier USS *Hornet* eight hours before it was able to pick up the capsule.

Part of this mission's payload: the Apollo Command Module 011. As planned, the associated Service Module was burned up on reentry into the earth's atmosphere. Splashdown took place in the Central Pacific, several hundred miles south of Wake Island.

SA 501 (APOLLO 4)

Saturn SA 501 was the fourteenth booster of the Saturn family and the first Saturn V. It was the first of a total of three unmanned Saturn Vs of the Apollo and Skylab programs. SA 501 delivered the Apollo Block 1 spacecraft CSM 017 and the Lunar Test Article 10R into a highly elliptical earth orbit.

Mission Data and Flight Information SA 501	
Official Mission Designation	Apollo 4
Saturn Mission Designation	Saturn Apollo (SA) 501
Payload	Apollo CSM-017 and LTA-10R
Launch Site	LC-39A
Payload Weight	81,262 pounds
Arrival of S-IC at the Cape	September 12, 1966
Arrival of S-II at the Cape	January 21, 1967
Arrival of S-IVB at the Cape	August 15, 1966
Instrument Unit at the Cape	May 20, 1967
Stacking S-IC	October 27, 1966
Stacking S-II	June 18, 1967
Stacking S-IVB	June 19, 1967
Stacking Instrument Unit	June 19, 1967
Stacking Apollo and LTA-10R	June 20, 1967
Rollout to Launch Pad 39A	August 26, 1967
Start of Countdown	November 4, 12:00
Start of Final Countdown	November 6, 1967, 22:30
S-IC Ignition Command	November 9, 11:59:52
Range Zero/Launch	12:00:01/000:00:00 MET
S-IC Cutoff Center Engine	000:02:16 MET
S-IC Cutoff Outer Engines	000:02:31 MET
S-II Engine Ignition	000:02:32 MET
Jettison S-II Interstage	000:03:01 MET
Escape Tower Separation	000:03:07 MET
S-II End of Burn	000:08:40 MET
S-IVB Ignition Start	000:08:41 MET
Start of First Burn Phase S-IVB	000:08:44 MET
S-IC Impact in Water	000:09:31 MET
Position S-IC	30°16'N, 74°35'W
End of First Burn S-IVB	000:11:06 MET
S-II Impact in Water	000:18:47 MET
Position of S-II	32°20'N, 39°83'W
Start of Second Burn Phase S-IVB	003:11:27 MET
End of Second Burn S-IVB	003:11:27 MET
CSM Separation from S-IVB	003:26:28 MET
S-IVB Impact (theoretical)	005:03:07 MET
Position S-IVB (theoretical)	23°43"N, 161°21'W

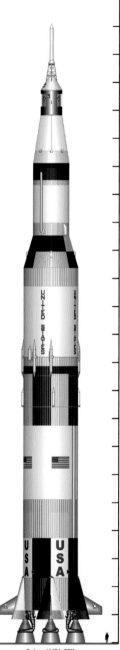

Saturn V (SA-501)
(NL : Apollo 4)

Assembly of SA 501 for the mission of Apollo 4 began on Friday, October 27, 1967, in High Bay 1 of the Vehicle Assembly Building (VAB). Because of delays in delivery of the S-II, a spacer had to be installed for function tests. This simulated the second stage and enabled the third stage to be attached. The rocket first had to be disassembled when the S-II arrived at the Cape. It was taken apart a second time when the S-II-1 had to undergo function tests again. As a result, final assembly could not take place until June 1967.

On August 26, 1967, Mobile Launcher 1 transported the rocket to Launch Complex 39A. Firing Room Number 1 in the launch control center was used to oversee the launch.

This mission was the first all up test flight. With the exception of the lunar lander, the booster was fitted with fully functional stages and systems. This was a deviation from the incremental tests previously used, in which just one new component was added for each mission. This meant that the Apollo 4 mission was the first flight of the S-IC first stage, the S-II second stage, the first flight of the Saturn V version of the S-IVB, and the first flight of the Saturn V instrument unit. The S-IVB used on the SA 501 mission was one of just two S-IVBs of the 500 series to reenter the atmosphere (the other was the S-IVB of Apollo 6).

The S-IC stages of SA 501 and SA 502 carried observation equipment which provided unique

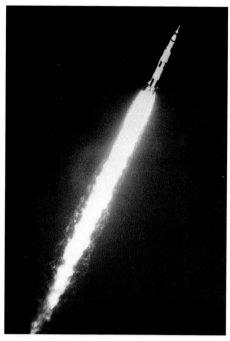

This photo was taken toward the end of Apollo 4's first minute of flight. It was taken by a flight tracking camera on the ground. The rocket was at an altitude of about 39,000 feet and was travelling at just below the speed of sound.

The S-II stage and the Apollo CSM 017 with the LTA-10R beneath the payload fairing in the VAB transfer area.

photographs. Two television cameras filmed the fiery surroundings of engine start and engine operation. The cameras were stowed above the heat shield at the base of the rocket. They could not have survived in the open area below. The camera lenses were oriented toward the engine area by a long fiber optic cable. They were also protected by quartz windows. Two cameras with fiber optics also made possible photographs from the interior of the oxygen tank, making it possible to observe the sloshing movements of the fuel during powered flight.

SA 204 (APOLLO 5)

Saturn SA 204 was the fifteenth booster of the Saturn family and the fourth Saturn IB. It was the fourth and last unmanned Saturn and it was the second and last flight by a Saturn IB which did not carry an Apollo Command and Service Module as payload. SA 204 delivered Lunar Module Number 1 to earth orbit for its first tests in orbit.

The booster rocket was the same one earmarked for the manned flight of Apollo 1 at Pad 34. Astronauts Grissom, White, and Chaffee lost their lives atop this rocket. The booster suffered no damage as a result of the Apollo fire, however. In March 1967, it was destacked at Pad 34 and one month later stacked at Pad 37B.

Stacking of SA 204 for the mission of Apollo 5 began on Wednesday, April 12, 1967, at Launch Complex 37B at the Cape Canaveral Air Force Base.

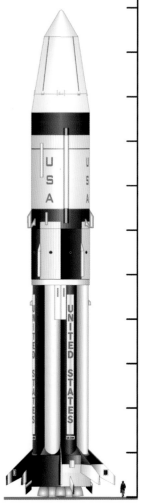

Saturn IB (SA-204)

Mission Data and Flight Information SA 204	
Official Mission Designation	Apollo 5
Saturn Mission Designation	Saturn Apollo (SA) 204
Payload	LM 1
Launch Site	LC-37B
Payload Weight	31,658 pounds
Arrival of S-IB at the Cape	August 15, 1966
Arrival of S-IVB at the Cape	August 6, 1966
Instrument Unit at the Cape	August 16, 1966
Stacking S-IB at Pad 37B	April 12, 1967
Stacking S-IVB	April 12, 1967
Stacking Instrument Unit	April 12, 1967
Stacking with LM 1	November 19, 1967
S-IB Ignition Command	January 22, 1968, 22:48:06
Range Zero/Launch	22:48:09/000:00:00 MET
S-IB Cutoff Center Engine	000:02:19 MET
S-IC Cutoff Outer Engines	000:02:22 MET
S-IB/S-IVB Separation Command	000:02:24 MET
End of Burn S-IVB	000:10:03 MET
Nosecone Jettison	000:10:44 MET
Jettison Adapter Panels	000:19:59 MET
Separation LM from S-IVB	000:53:50 MET
S-IV Reentry	015:32:00 MET
S-IVB Wiedereintritt	015:32:00 MET

This photo was taken on August 2, 1967, and shows the payload fairing nosecone of the Saturn IB used on Mission SA 204.

As the Saturn IB flew this mission without a crew, use of an escape tower was not necessary. The Lunar Module was stowed under a cone-shaped payload fairing. The height of the complete booster was therefore just 180 feet (instead of the usual 223 feet).

The Saturn IB delivered the LM 1 to an orbit with a perigee of 101 miles and an apogee of 138 miles. The rocket's nose cone was blown off after 43 minutes and 52 seconds, then the Lunar Module was released.

Technicians at the Cape Canaveral Air Base have just hoisted the payload adapter with Lunar Module 1 into position, ready to be mounted on SA 204.

SA 502 (APOLLO 6)

Saturn SA 502 was the sixteenth booster of the Saturn family and the second Saturn V. It was the second unmanned test of the Saturn V. The payload consisted of Apollo Command Module No. 020, Apollo Service Module No. 014, and Lunar Test Article 2R, a mockup of the lunar lander.

Mission Data and Flight Information SA 502	
Official Mission Designation	Apollo 6
Saturn Mission Designation	Apollo Saturn (SA) 502
Payload	CM 020, SM 014, and LTA 2R
Launch Site	LC-39A
Payload Weight	55,424 and 25,992 pounds
Arrival of S-IC at the Cape	March 3, 1967
Arrival of S-II at the Cape	May 24, 1967
Arrival of S-IVB at the Cape	July 14, 1967
Instrument Unit at the Cape	March 20, 1967
Stacking S-IC	March 29, 1967
Stacking S-II	July 13, 1967
Stacking S-IVB	July 14, 1967
Stacking Instrument Unit	July 14, 1967
Stacking Apollo and LTA-2R	December 10, 1967
Rollout to Launch Pad 39B	February 6, 1968
Start of Final Countdown	April 4, 1968, 08:00
S-IC Ignition Command	April 4, 1968, 11:59:51
Range Zero/Launch	12:00:00/000:00:00 MET
Loss of Adapter Parts	000:02:13 MET
S-IC Cutoff Center Engine	000:02:15 MET
S-IC Cutoff Outer Engines	000:02:42 MET
S-II Engine Ignition	000:02:44 MET
Jettison S-II Interstage	000:03:12 MET
Escape Tower Separation	000:03:18 MET
Engine 2 Power Loss	000:03:45 MET
Engine 2 Shutdown	000:06:52 MET
Engine 3 Shutdown	000:06:55 MET
S-II Cutoff Engines 1, 4 & 5	000:10:11 MET
Present Position S-IC	30°12′N, 74°19′W
S-IVB Ignition Start	000:10:12 MET
Start of First Burn Phase S-IVB	000:10:15 MET
S-IC Impact in Water	000:09:31 MET
Position S-IC	30°16′N, 74°35′W
End of First Burn S-IVB	000:13:11 MET
Position S-II	31°12′N, 32°1′W
Start of Second Burn Phase S-IVB	003:11:27 MET
Attempted Second Burn S-IVB	003:13:00 MET
CSM Separation from S-IVB	003:14:42 MET
S-IVB Reentry into Earth's Atmosphere	April 25, 1968

The launch of Apollo 6 at 0700 local time, on April 4, 1968.

The Instrument Unit is installed on the Apollo 6 S-IVB stage.

pogo oscillations, which caused damage to parts of the rocket. During the flight of the first stage a panel of the payload fairing separated because of the vibration and expanding moisture in the structure. During operation of the second stage, breaks in the igniter fuel lines caused two of the five engines to shut down prematurely, causing the third stage and payload to achieve a suboptimal orbit. Instead of the desired orbit at 118 miles, an elliptical orbit with a perigee of 107 miles and an apogee of 223 miles was achieved. Reignition of the third stage also failed, causing the S-IVB to remain in a low orbit. A few weeks after launch, it reentered the earth's atmosphere and burned up.

During missions SA 501 and SA 502—in addition to the cameras in the engine area of the S-IC stage, which filmed the burn behavior of the F-1 engines—two other cameras captured the spectacular moments of separation of the S-IC from the S-II stage. Immediately after separation of the interstage from the S-II stage they were jettisoned in a watertight capsule and came down by parachute far out in the Atlantic, where they were retrieved.

Stacking of SA 502 for the Apollo 6 mission began on Wednesday, March 29, 1967, in High Bay 3 of the Vehicle Assembly Building. Stacking up to the instrument unit had to be carried out twice, because delivery of the S-II was delayed. So that the combination could at least be tested without the S-II, it was equipped with a dumbbell-shaped spacer which simulated the weight and dimensions of the second stage and also had all the electrical fittings. Ultimate assembly up to the instrument unit did not take place until mid-July.

On September 29, 1967, Apollo Block 1 Command Module 014, which was combined with Service Module 020, arrived at the Cape. With Lunar Test Article No. 2 it was integrated into the payload fairing and on December 10 was fitted to the rocket. On February 6, 1968, Mobile Launcher 2 transported SA 502 to Launch Pad 39A.

Firing Room Number 2 in the launch control building at the Cape was used for this launch.

Unlike the nearly perfect launch of the first Saturn V, this mission ran into serious difficulties right from the beginning. The reason for this was

Thirty seconds after ignition of the S-II stage's engines, the interstage, the connecting element between the S-IC and S-II stages, is jettisoned.

SA 205 (APOLLO 7)

Saturn SA 205 was the seventeenth booster of the Saturn family and the fifth Saturn IB. It was the very first manned flight of a Saturn IB. AS 205 delivered astronauts Walter Schirra, Donn Eisele, and Walter Cunningham into earth orbit on the first manned mission of the Apollo program.

Stacking of SA 205 for the Apollo 7 mission began on Friday, August 9, 1968, at Launch Pad 34. Apollo 7 was the only flight of the entire manned spaceflight program after Gemini which did not begin from the Kennedy Space Center. All American manned spaceflights after Apollo 7 either began on Launch Pad 38A or 39 B of the Kennedy Space Center.

Mission Data and Flight Information SA 205	
Official Mission Designation	Apollo 7
Saturn Mission Designation	Apollo Saturn (SA) 205
Payload	Apollo CSM 101
Launch Site	LC-34
Payload Weight	36,420 pounds
Arrival of S-IB at the Cape	March 28, 1968
Arrival of S-IVB at the Cape	April 8, 1968
Instrument Unit at the Cape	April 11, 1968
Stacking S-IB	August 9, 1968
Stacking S-IVB	August 9, 1968
Stacking Instrument Unit	August 9, 1968
Stacking with Apollo CSM	August 30, 1968
Start of Countdown	October 6, 1968, 19:00
Start of Final Countdown	October 10, 1968, 14:30
S-IC Ignition Command	October 11, 1968, 15:02:42
Range Zero/Launch	15:02:45/000:00:00 MET
S-IB Cutoff Center Engine	000:02:21 MET
S-IC Cutoff Outer Engines	000:02:24 MET
S-IB/S-IVB Separation Command	000:02:26 MET
S-IVB Ignition Command	000:02:27 MET
Escape Tower Separation	000:02:47 MET
S-IB Impact with Water	000:09:20 MET
S-IVB End of Burn	000:10:17 MET
End of Passivation Experiments	00:11:15 MET
Last Sighting of CSM	053:20:00 MET
S-IVB Reentry	162:27:15 MET
Ende Passivierungsexperimente	005:11:15 MET
Letzte Sichtung von CSM aus	053:20:00 MET
S-IVB Wiedereintritt	162:27:15 MET

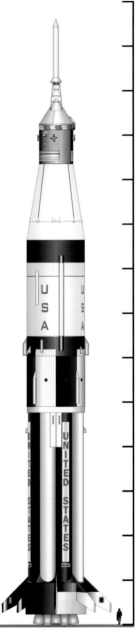

Saturn IB
(SA-205)

The second stage (S-IVB) of Saturn 205, the booster rocket used on the first manned mission of the Apollo program, arrives at Cape Canaveral Air Force Base on board the Super Guppy of Aero Spacelines, on Monday, April 8, 1968.

After the insertion of Apollo 7 into earth orbit, the S-IVB stage was used for extensive passivation experiments. They were carried out in seven stages, beginning twenty seconds after reaching orbit and continuing for the next five hours.

The separation of the Apollo Command and Service Module was not carried out until two hours and fifty-five minutes into the mission, to simulate the separation events that would occur during a flight to the moon. In the process, the astronauts discovered that the four panels of the payload fairing had not completely opened. This had no effect on the rest of the mission, however. Later flights of the S-IVB during Saturn V missions saw the panels blown off.

After the separation, Apollo 7 carried out rendezvous maneuvers and tracking experiments for a period of more than two days. Apollo 7's last sighting of the S-IVB occurred at 053:20:00 MET from a distance of 367 miles.

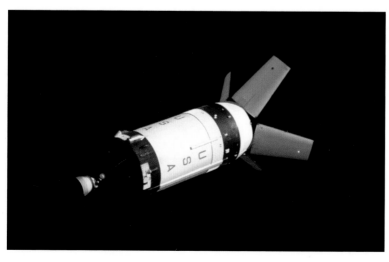

The S-IVB stage with open adapter panels in space. The photo was taken by Apollo 7 astronauts Schirra, Cunningham, and Eisele soon after they reached orbit.

SA 503 (APOLLO 8)

S A 503 was the eighteenth booster of the Saturn family and the third Saturn V. It was the second manned Saturn of the Apollo

Mission Data and Flight Information SA 503	
Official Mission Designation	Apollo 8
Saturn Mission Designation	Apollo Saturn (SA) 503
Payload	Apollo CSM 103 and LTA
Launch Site	LC-39A
Payload Weight	63,647 and 19,841 pounds
Arrival of S-IC at the Cape	December 27, 1967
Arrival of S-II at the Cape	June 27, 1968
Arrival of S-IVB at the Cape	December 30, 1967
Instrument Unit at the Cape	January 4, 1968
Stacking S-IC	December 30, 1967
Stacking S-II	June 27, 1968
Stacking S-IVB	August 14, 1968
Stacking Instrument Unit	August 15, 1968
Stacking Apollo and LTA	October 7, 1968
Rollout to Launch Pad 39A	October 9, 1968
Start of Countdown	December 16, 00:00
Start of Final Countdown	December 20, 01:51
S-IC Ignition Command	December 21, 1968, 12:50:51
Range Zero/Launch	12:51:00/000:00:00 MET
S-IC Cutoff Center Engine	000:02:06 MET
S-IC Cutoff Outer Engines	000:02:34 MET
S-II Engine Ignition	000:02:36 MET
Jettison S-II Interstage	000:03:04 MET
Escape Tower Separation	000:03:09 MET
S-II End of Burn	000:08:44 MET
S-IVB Ignition Start	000:08:45 MET
Start of First Burn Phase S-IVB	000:08:48 MET
S-IC Impact in Water	000:09:04 MET
Position S-IC	30°12′N, 74°7′W
End of First Burn S-IVB	000:11:25 MET
Position S-II	31°50′N, 38°0′W
Start of Second Burn Phase S-IVB	002:50:38 MET
CSM Separation from S-IVB	003:20:59 MET
CSM Spacing Maneuver from S-IVB	003:40:01 MET
Ignition APS	005:25:56 MET
APS End of Burn	005:38:09 MET
Closest Approach to Moon	069:58:55 MET
Present Location of S-IVB 503	Heliocentric Orbit
Größte Annäherung an Mond	069:58:55 MET
Heutiger Verbleib S-IVB 503	Heliozentrische Umlaufbahn

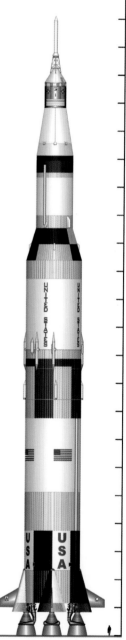

Saturn V (SA-503)
(NL : Apollo 8)

The S-IC stage of Saturn V SA 503 is prepared for final assembly of the rocket in the High Bay of the Vehicle Assembly Building at Kennedy Space Center.

program and the first manned flight of a Saturn V. SA 503 sent astronauts Frank Borman, James Lovell, and William Anders on their way to the first circuit of the moon.

Assembly of SA 503 for the Apollo 8 mission began on Saturday, December 30, 1967, in High Bay 1 of the Vehicle Assembly Building. When, at the beginning of 1968, it became clear that this Saturn V would be used for a manned flight, it was unstacked so that the S-II could be transported back to the Manned Spaceflight Center. There modifications were made to prevent a recurrence of the pogo oscillations encountered by Apollo 6. Finally assembly began again in July.

On October 7, the stack with the Apollo spacecraft CSM 103 and the Lunar Test Article (LTA), a dummy simulating the weight of the Lunar Module, which would be carried on this mission, was attached to the rocket. On October 9, 1968, Mobile Launcher 1 transported the rocket to Launch Complex 39A.

Firing Room Number 1 in the launch control center was chosen to monitor the launch.

Nine minutes after leaving the launch pad, the burnt-out first stage crashed into the Atlantic approximately 350 miles off the coast of Florida. The four ullage solid-fuel engines, together with their fairings, were separated at 000:08:57 MET.

The S-IVB used in mission SA 503 bore the serial number S-IVB-503N. The N stood for New, for the stage originally earmarked for this flight was destroyed in a test.

A scene from Firing Room No.1 at the Kennedy Space Center, on December 21, 1968, a few hours before the launch of Apollo 8 on its historic mission.

SA 504 (APOLLO 9)

S A 504 was the nineteenth booster of the Saturn family and the fourth Saturn V. It was the third manned Saturn of the Apollo program and the second manned flight of a Saturn V.

Mission Data and Flight Information SA 504	
Official Mission Designation	Apollo 9
Saturn Mission Designation	Apollo Saturn (SA) 504
Payload	Apollo CSM 104 and LM 3
Launch Site	LC-39A
Payload Weight	63,493 and 32,132 pounds
Arrival of S-IC at the Cape	September 30, 1968
Arrival of S-II at the Cape	May 15, 1968
Arrival of S-IVB at the Cape	September 12, 1968
Instrument Unit at the Cape	September 30, 1968
Stacking S-IC	October 1, 1968
Stacking S-II	October 3, 1968
Stacking S-IVB	October 6, 1968
Stacking Instrument Unit	October 7, 1968
Stacking Apollo and LM	December 3, 1968
Rollout to Launch Pad 39A	January 3, 1969
Start of Final Countdown	February 27, 1969, 03:00
S-IC Ignition Command	March 3, 1969, 15:59:51
Range Zero/Launch	16:00:00/000:00:00 MET
S-IC Cutoff Center Engine	000:02:14 MET
S-IC Cutoff Outer Engines	000:02:43 MET
S-II Engine Ignition	000:02:45 MET
Jettison S-II Interstage	000:03:14 MET
Escape Tower Separation	000:03:18 MET
S-IC Impact in Water	000:08:56 MET
Position S-IC	30°11'N, 74°14'W
S-II Cutoff	000:08:56 MET
S-IVB Ignition Start	000:09:01 MET
Start of First Burn Phase S-IVB	000:09:09 MET
End of First Burn S-IVB	000:11:05 MET
S-II Impact in Water	000:20:25 MET
Position S-II	31°28'N, 34°02'W
CSM Separation from S-IVB	002:41:16 MET
CSM Docking with LM/S-IVB	003:01:59 MET
CSM/LM Separation from S-IVB	004:08:09 MET
Start of Second Burn Phase S-IVB	004:45:47 MET
End of Second Burn Phase S-IVB	004:46:58 MET
Start of Third Burn Phase S-IVB	006:07:19 MET
End of Third Burn Phase S-IVB	006:11:21 MET
Ignition APS	007:34:05 MET
APS End of Burn	007:41:53 MET
Present Location of S-IVB 504	Heliocentric Orbit

Launch of Apollo 9 at 1100 local time, on March 3, 1969.

This photo shows the Apollo 9 S-IC stage built in Michoud on its way to the dock where it was loaded onto the barge for shipment to the Mississippi Test Facility for acceptance tests.

SA 504 put astronauts James McDivitt, David Scott, and Russell Schweickart into earth orbit to test the interaction between the Lunar Module and the Apollo command ship.

The assembly of SA 504 for the Apollo 9 mission began on Tuesday, October 1, 1968, in High Bay 3 of the Vehicle Assembly Building. On January 3, 1969, Mobile Launcher 2 moved the rocket to Launch Pad 39A, three miles from the Vehicle Assembly Building.

Firing Room Number 2 in the launch control building at the Cape was used for this launch.

The S-IVB's ullage case was separated at 000:09:09 MET.

For the Saturn V used on Apollo 9 it was a unique flight in the Apollo program. First, it was the only Saturn V ever to deliver its crew into earth orbit. It was also the only time that the third stage's J-2 engine was fired three times. Two of the ignitions took place after the Apollo 9 command ship and the attached Lunar Module had separated. Ignition number two, with a duration of seventy seconds, placed the stage on a medium-level orbit, while ignition three finally placed it in a heliocentric orbit, where it remains to this day.

At the end of the mission the booster rocket's Auxiliary Propulsion System (APS) was ignited and fired until its fuel was consumed. It ran for almost seven minutes. This was by far the longest APS burn phase of any Apollo mission. Passivation of the S-IVB failed, however, because a problem developed in the valve system. As a result it was not possible to discharge the remaining hydrogen and oxygen. This had no negative effects, at least while telemetry could follow the stage. An explosion in a later phase of flight, outside the range of observation, was possible however.

SA 504's S-IVB stage in earth orbit. The adapter panels have been jettisoned. The Apollo 9 Command and Service Module was now able to dock with the Lunar Module and extract it from the adapter area.

SA 505 (APOLLO 10)

S A 505 was the twentieth booster of the Saturn family and
the fifth Saturn V. It was the fourth manned Saturn of the

Launch of Apollo 10, on May 18, 1969.

Mission Data and Flight Information SA 505	
Official Mission Designation	Apollo 10
Saturn Mission Designation	Apollo Saturn (SA) 505
Payload	Apollo CSM 106 and LM 4
Launch Site	LC-39B
Payload Weight	63,570 and 32,732 pounds
Arrival of S-IC at the Cape	November 27, 1968
Arrival of S-II at the Cape	December 3, 1968
Arrival of S-IVB at the Cape	December 10, 1968
Instrument Unit at the Cape	December 15, 1968
Stacking S-IC	November 30, 1968
Stacking S-II	December 7, 1968
Stacking S-IVB	December 29, 1968
Stacking Instrument Unit	December 30, 1968
Stacking Apollo and LM	February 6, 1969
Rollout to Launch Pad 39B	March 11, 1969
Start of Final Countdown	May 17, 1969, 01:00
S-IC Ignition Command	March 18, 1969, 16:48:51
Range Zero/Launch	16:49:00/000:00:00 MET
S-IC Cutoff Center Engine	000:02:15 MET
S-IC Cutoff Outer Engines	000:02:42 MET
S-II Engine Ignition	000:02:44 MET
Jettison S-II Interstage	000:03:12 MET
Escape Tower Separation	000:03:18 MET
S-IC Impact in Water	000:08:59 MET
Position S-IC	29°42'N, 73°39'W
S-II Outer Engine Cutoff	000:09:13 MET
S-IVB Ignition Start	000:09:14 MET
Start of First Burn Phase S-IVB	000:09:17 MET
End of First Burn S-IVB	000:11:05 MET
S-II Impact in Water	000:20:18 MET
Position S-II	31°31'N, 34°31'W
Start of Second Burn Phase S-IVB	002:33:28 MET
End of Second Burn Phase S-IVB	002:39:11 MET
CSM Separation from S-IVB	003:02:42 MET
CSM Docking with LM/S-IVB	003:17:36 MET
CSM/LM Separation from S-IVB	003:56:26 MET
Ignition APS for Separation Maneuver	005:28:56 MET
APS End of Burn	005:29:05 MET
Closest Approach to Moon	078:51:03 MET
Present Location of S-IVB 505	Heliocentric Orbit

Tugs deliver the second stage for Mission SA 505 (Apollo 10) to the test stand at the Mississippi Test Facility (maintenance test flight). Transport between Michoud the maintenance test flight was carried out with uncovered barges.

Apollo program and the third manned flight of a Saturn V.

SA 505 put astronauts Thomas Stafford, John Young, and Eugene Cernan on the way to the moon for a general test of a moon landing.

The assembly of SA 505 for the Apollo 10 mission began on Thursday, November 30, 1968, in High Bay 2 of the Vehicle Assembly Building. On February 6, the combination of Apollo spacecraft CSM 106 and Lunar Module 4, already integrated in the payload adapter, were attached to the rocket. On March 11, 1969, Mobile Launcher 3 moved the rocket to Launch Pad 39B, four miles from the Vehicle Assembly Building.

Firing Room Number 3 in the launch control building at the Cape was used for this launch.

The ullage case was separated at 000:09:21 MET.

The S-IVB's auxiliary propulsion system (APS) consisted of two identical units on opposite sides of the rocket at the base of the S-IVB stage. Each APS consisted of four thrusters: a single Rocketdyne SE 7-1 from the Gemini program with a thrust of 70 pounds for forward acceleration and three TR-204-

150 thrusters each producing 150 pounds of thrust. These three thrusters were used for attitude control. The Panther-204 was possibly the most reliable power plant of the entire booster, for it had a quadruple redundant valve system. Even the valve system in the lunar lander's ascent engine was only designed with double redundancy.

The first stage of Apollo 10 arrives by barge at the Kennedy Space Center on November 27, 1968, and is taken to the Vehicle Assembly Building (in the background) for final assembly.

SA 506 (APOLLO 11)

S A 506 was the twenty-first booster of the Saturn family and the sixth Saturn V. It was the fifth manned Saturn of the Apollo program

Mission Data and Flight Information SA 506	
Official Mission Designation	Apollo 11
Saturn Mission Designation	Apollo Saturn (SA) 506
Payload	Apollo CSM 107 and LM 5
Launch Site	LC-39A
Payload Weight	63,504 and 33,279 pounds
Arrival of S-IC at the Cape	February 20, 1969
Arrival of S-II at the Cape	February 6, 1969
Arrival of S-IVB at the Cape	January 19, 1969
Instrument Unit at the Cape	February 27, 1969
Stacking S-IC	February 27, 1969
Stacking S-II	March 4, 1969
Stacking S-IVB	March 5, 1969
Stacking Instrument Unit	March 5, 1969
Stacking Apollo and LM	April 14, 1969
Rollout to Launch Pad 39A	May 20, 1969
Start of Final Countdown	July 14, 1969, 21:00
S-IC Ignition Command	July 16, 1969, 13:31:51
Range Zero/Launch	13:32:00/000:00:00 MET
S-IC Cutoff Center Engine	000:02:15 MET
S-IC Cutoff Outer Engines	000:02:42 MET
S-II Engine Ignition	000:02:44 MET
Jettison S-II Interstage	000:03:12 MET
Escape Tower Separation	000:03:18 MET
S-IC Impact in Water	000:09:04 MET
Position S-IC	30°13'N, 74°2'W
S-II Center Engine Cutoff	000:09:08 MET
S-IVB Ignition Start	000:09:09 MET
Start of First Burn Phase S-IVB	000:09:12 MET
End of First Burn S-IVB	000:11:39 MET
S-II Impact in Water	000:20:14 MET
Position S-II	31°32'N, 34°51'W
Start of Second Burn Phase S-IVB	002:44:16 MET
End of Second Burn Phase S-IVB	002:50:03 MET
CSM Separation from S-IVB	003:15:23 MET
CSM Docking with LM/S-IVB	003:24:04 MET
CSM/LM Separation from S-IVB	004:17:03 MET
Ignition APS for Separation Maneuver	005:37:48 MET
APS End of Burn	005:42:28 MET
Closest Approach to Moon	078:42:00 MET
Present Location of S-IVB 506	Heliocentric Orbit
Heutiger Verbleib S-IVB 506	Heliozentrische Umlaufbahn

The launch of SA 506 with Apollo 11 on the historic first manned landing on the moon.

February 21, 1969: a crane brings SA 506's S-IC into position for stacking, which began a few days later.

and the fourth manned flight of a Saturn V.

SA 506 put astronauts Neil Armstrong, Edwin "Buzz" Aldrin, and Michael Collins on their way to the historic first landing on the moon.

The assembly of SA 506 for the Apollo 11 mission began on Thursday, February 27, 1969, in High Bay 1 of the Vehicle Assembly Building and was completed one week later. The stack with Apollo spacecraft CSM 107 and Lunar Module 5 arrived at the beginning of March and was mounted on the tip of the rocket. On May 20, 1969, Mobile Launcher 1 moved the rocket to Launch Pad 39A, three miles from the Vehicle Assembly Building.

Firing Room Number 1 in the launch control building at the Cape was used for this launch.

About nine minutes after launch, the burnt-out first stage crashed into the Atlantic about 235 miles off the coast of Florida. The four solid-fuel ullage motors ignited one second before the S-IC's four outer engines shut down. Their purpose was to provide sufficient forward acceleration in the phase between the first stage end of burn and the buildup of thrust by the second stage to ensure that the fuel remained at the base of the tank and no bubbles formed in the lines. The entire burn time of these solid-fuel TX-280 motors was 3.5 seconds. Together with their casings, they were separated at 000:09:21 MET.

SA 506's S-IVB stage is attached to the S-II stage, on March 5, 1969.

SA 507 (APOLLO 12)

S A 507 was the twenty-second booster of the Saturn family and the seventh Saturn V. It was the sixth manned Saturn of the

Mission Data and Flight Information SA 507	
Official Mission Designation	Apollo 12
Saturn Mission Designation	Apollo Saturn (SA) 507
Payload	Apollo CSM 108 and LM 6
Launch Site	LC-39A
Payload Weight	63,581 and 33,587 pounds
Arrival of S-IC at the Cape	May 3, 1969
Arrival of S-II at the Cape	April 21, 1969
Arrival of S-IVB at the Cape	March 10, 1969
Instrument Unit at the Cape	May 8, 1969
Stacking S-IC	May 7, 1969
Stacking S-II	May 21, 1969
Stacking S-IVB	May 22, 1969
Stacking Instrument Unit	May 22, 1969
Stacking Apollo and LM	July 1, 1969
Rollout to Launch Pad 39A	September 8, 1969
Start of Final Countdown	November 13, 1969, 02:00
S-IC Ignition Command	November 14, 1969, 16:21:51
Range Zero/Launch	16:22:00/000:00:00 MET
S-IC Cutoff Center Engine	000:02:15 MET
S-IC Cutoff Outer Engines	000:02:42 MET
S-II Engine Ignition	000:02:43 MET
Jettison S-II Interstage	000:03:12 MET
Escape Tower Separation	000:03:18 MET
S-II Center Engine Cutoff	000:07:41 MET
S-IC Impact in Water	000:09:15 MET
Position S-IC	30°16'N, 74°54'W
S-II Outer Engine Cutoff	000:09:12 MET
S-IVB Ignition Start	000:09:13 MET
Start of First Burn Phase S-IVB	000:09:17 MET
End of First Burn S-IVB	000:11:34 MET
S-II Impact in Water	000:20:22 MET
Position S-II	31°28'N, 34°13'W
Start of Second Burn Phase S-IVB	002:44:23 MET
End of Second Burn Phase S-IVB	002:53:04 MET
CSM Separation from S-IVB	003:18:05 MET
CSM Docking with LM/S-IVB	003:26:53 MET
CSM/LM Separation from S-IVB	004:13:00 MET
Ignition APS for Separation Maneuver	005:23:20 & 005:29:13 MET
APS End of Burn	005:28:20 & 005:33:40 MET
Closest Approach to Moon	085:48:00 MET
Present Location of S-IVB 506	Heliocentric Orbit

The launch of Apollo 12 at 1122 Eastern Time in extremely bad weather. A few seconds later the rocket was struck by lightning.

Apollo program and the fifth manned flight of a Saturn V.

SA 506 put astronauts Charles Conrad, Richard Gordon, and Alan Bean on their way to the second landing on the moon.

The assembly of SA 507 for the Apollo 12 mission began on Wednesday, May 7, 1969, in High Bay 3 of the Vehicle Assembly Building and was completed one week later. Apollo spacecraft CSM 108 and Lunar Module 6 were mounted on the tip of the rocket on July 1. On September 8, 1969, Mobile Launcher 2 moved the rocket to Launch Pad 39A.

Firing Room Number 2 in the launch control building at the Cape

SA 507's S-IC stage is prepared for stacking, on May 7, 1969.

was used for this launch. As usual, the countdown began ninety-eight hours before launch and the final countdown twenty-eight hours before liftoff at 0200 Greenwich Mean Time (GMT) on November 13, 1969.

The flight began with a thunderclap. During the initial phase of the ascent the rocket was hit twice by lightning—once at thirty-six flying seconds, the other time at fifty-two seconds. Many of the instruments on the Apollo and the main power supply were temporarily knocked out, but the Saturn V's instrumentation and all mechanical and pyrotechnic components functioned normally, allowing the threatened flight abort to be avoided.

The casings of the ullage motors were separated at 000:09:25 MET.

The APS burn maneuver to conclude the Saturn V mission was carried out in two steps this time. After separation of the Lunar Module it was planned that the S-IVB would set course for a solar orbit. The APS was supposed to guide it past the moon's trailing edge. The Apollo spacecraft always approached the moon from the other side. The moon's gravity was then to send the S-IVB into a solar orbit in a swing by maneuver. A small error in the state vector in the Saturn's guidance system caused the S-IVB to fly past the Moon at too high an altitude to achieve Earth escape velocity. Unable to achieve a heliocentric orbit, the stage remained

in an unstable, highly-elliptical earth orbit until in 1971, after another close flypast of the moon, it finally reached a solar orbit. After that it disappeared, until on September 3, 2003, an amateur astronomer discovered it, first believing it to be a near-earth asteroid. It was initially given the designation J002E3 before it was determined that it was in fact an artificial object.

The S-II stage is placed on the S-IC on May 21.

SA 508 (APOLLO 13)

S A 508 was the twenty-third booster of the Saturn family and the eighth Saturn V. It was the seventh manned Saturn of the

Mission Data and Flight Information SA 508	
Official Mission Designation	Apollo 13
Saturn Mission Designation	Apollo Saturn (SA) 508
Payload	Apollo CSM 109 and LM 7
Launch Site	LC-39A
Payload Weight	63,471 and 33,510 pounds
Arrival of S-IC at the Cape	June 16, 1969
Arrival of S-II at the Cape	June 29, 1969
Arrival of S-IVB at the Cape	June 13, 1969
Instrument Unit at the Cape	July 7, 1969
Stacking S-IC	June 18, 1969
Stacking S-II	July 17, 1969
Stacking S-IVB	July 31, 1969
Stacking Instrument Unit	August 1, 1969
Stacking Apollo and LM	December 10, 1969
Rollout to Launch Pad 39A	December 15, 1969
Start of Final Countdown	April 10, 1970, 05:00
S-IC Ignition Command	April 11, 19:12:51
Range Zero/Launch	19:13:00/000:00:00 MET
S-IC Cutoff Center Engine	000:02:15 MET
S-IC Cutoff Outer Engines	000:02:44 MET
S-II Engine Ignition	000:02:46 MET
Jettison S-II Interstage	000:03:14 MET
Escape Tower Separation	000:03:21 MET
S-II Center Engine Cutoff	000:05:31 MET
S-IC Impact in Water	000:09:07 MET
Position S-IC	30°11′N, 74°04′W
S-II Outer Engine Cutoff	000:09:53 MET
S-IVB Ignition Start	000:09:54 MET
Start of First Burn Phase S-IVB	000:09:57 MET
End of First Burn S-IVB	000:12:30 MET
S-II Impact in Water	000:20:58 MET
Position S-II	32°19′N, 33°17′W
Start of Second Burn Phase S-IVB	002:35:46 MET
End of Second Burn Phase S-IVB	002:41:37 MET
CSM Separation from S-IVB	003:06:39 MET
CSM Docking with LM/S-IVB	003:19:09 MET
CSM/LM Separation from S-IVB	004:01:01 MET
Ignition APS for Lunar Impact	006:00:00 MET
APS End of Burn	006:03:37 MET
S-IVB Impact on Moon	077:56:40 MET
Present Location of S-IVB 507	Moon, Fra Mauro Region

The launch of SA 509 on the Apollo 13 mission, at 1313 Houston Time. It seemed a bad omen to contemporary observers. At Cape Canaveral, the launch site, it was only 1213 local time.

Apollo program and the sixth manned flight of a Saturn V.

SA 506 put astronauts James Lovell, John Swigert, and Fred Haise on the way to the planned third landing on the moon, however the mission had to be aborted because of severe damage in the Service Module of CSM 109.

The assembly of SA 508 for the Apollo 13 mission began on Wednesday, June 18, 1969, in High Bay 1 of the Vehicle Assembly Building and was completed one week later. Apollo spacecraft CSM 109 and Lunar Module 7 were mounted on the tip of the rocket on December 10. On December 15,

Definitely not on its way to the launch pad. On August 8, 1969, SA 508 with a Boilerplate spacecraft was driven once around the block when it moved from High Bay 2 to High Bay 3 and had to leave the Vehicle Assembly Building.

1969, Mobile Launcher 3 moved the rocket to Launch Pad 39A. Firing Room Number 1 in the launch control building at the Cape was used for this launch.

The mission was ill-fated from the very beginning. Just three days before the launch pilot Ken Mattingly had to be replaced. Charlie Duke of the replacement crew had come into contact with measles and then with Mattingly. The flight surgeons determined that Mattingly was the only member of the crew who had no antibodies in his blood.

During the launch itself there were powerful pogo oscillations, which shut down the center engine of the S-II stage more than two minutes too early. The outer engines had to run thirty-five seconds longer than planned to partially compensate for the loss of thrust. The ullage case was separated at 000:10:05 MET.

The S-IVB's first burn stage had to be extended by nine seconds to eliminate the remaining velocity deficit.

The S-IVB's auxiliary propulsion system was used twice after the separation of the Command and Service Module and Lunar Module. The first was between 004:18:01 MET and 004:19:21 MET, to separate from the spacecraft. The second maneuver between 006:00:00 MET and 006:03:37 MET placed the S-IVB on a course which led to it impacting with the moon's surface at 077:56:40 MET not far from the Apollo 12 landing site, near the Fra Mauro Crater. This was one of the few aspects of this mission which went according to plan.

The third stage adapter is lowered over Lunar Module Aquarius, whose supplies would later save the lives of the three astronauts.

SA 509 (APOLLO 14)

SA 509 was the twenty-fourth booster of the Saturn family and the ninth Saturn V. It was the eighth manned Saturn of the Apollo

Mission Data and Flight Information SA 509	
Official Mission Designation	Apollo 14
Saturn Mission Designation	Apollo Saturn (SA) 509
Payload	Apollo CSM 110 and LM 8
Launch Site	LC-39A
Payload Weight	64,463 and 33,653 pounds
Arrival of S-IC at the Cape	January 11, 1970
Arrival of S-II at the Cape	January 21, 1970
Arrival of S-IVB at the Cape	January 20, 1969
Instrument Unit at the Cape	May 6, 1970
Stacking S-IC	January 14, 1970
Stacking S-II	May 12, 1970
Stacking S-IVB	May 13, 1970
Stacking Instrument Unit	May 14, 1970
Stacking Apollo and LM	November 4, 1970
Rollout to Launch Pad 39A	November 9, 1970
Start of Final Countdown	January 30, 1971, 06:00
S-IC Ignition Command	January 31, 1971, 21:02:53
Range Zero/Launch	21:03:02/000:00:00 MET
S-IC Cutoff Center Eengine	000:02:15 MET
S-IC Cutoff Outer Engines	000:02:44 MET
S-II Engine Ignition	000:02:47 MET
Jettison S-II Interstage	000:03:15 MET
Escape Tower Separation	000:03:21 MET
S-II Center Engine Cutoff	000:07:43 MET
S-IC Impact in Water	000:09:06 MET
Position S-IC	29°50'N, 74°03'W
S-II Outer Engine Cutoff	000:09:19 MET
S-IVB Ignition Start	000:09:20 MET
Start of First Burn Phase S-IVB	000:09:23 MET
End of First Burn S-IVB	000:11:41 MET
S-II Impact in Water	000:20:46 MET
Start of Second Burn Phase S-IVB	002:28:32 MET
End of Second Burn Phase S-IVB	002:34:23 MET
CSM Separation from S-IVB	003:02:29 MET
CSM Docking with LM/S-IVB	004:56:57 MET
CSM/LM Separation from S-IVB	005:47:14 MET
Ignition APS for Lunar Impact	008:59:59 MET
APS End of Burn	009:04:11 MET
S-IVB Impact on the Moon	082:37:53 MET
Lunar Coordinates S-IVB 509	8°09'S, 26°02'W
Mond-Koordinaten S-IVB 509	8°09'S, 26°02'W

The launch of SA 509 photographed by a camera on the Service Tower of Launch Complex 39A.

SA 509's S-II stage being unloaded in the so-called "Turn Basin" at Kennedy Space Center after arriving by barge.

program and the seventh manned flight of a Saturn V.

SA 509 put astronauts Alan Shepard, Stuart Roosa, and Edgar Mitchell on the way to the third landing on the moon.

The assembly of SA 509 for the Apollo 14 mission began on Wednesday, January 14, 1970, in High Bay 3 of the Vehicle Assembly Building. Apollo spacecraft CSM 110 and Lunar Module 8 were mounted on the tip of the rocket on November 4, 1970. On November 9, Mobile Launcher 2 moved the rocket to Launch Pad 39A. Firing Room Number 2 in the launch control building at the Cape was used for this launch.

The ullage case was separated at 000:09:32 MET. Shortly after launch, docking of Kitty Hawk with the Antares caused major problems. Stuart Roosa needed six attempts to dock the CSM with the LM, because the docking latches failed to engage the Lunar Module's docking tunnel. The cause for this was never discovered. At the time

a foreign object in the mechanism was suspected. The crew of the Apollo spacecraft had to come up with a special strategy to complete the docking. They decided to fire the Service Module's thrusters for several seconds after the sixth contact with the mechanism, pressing the Apollo against the Lunar Module. Finally, the clamps engaged. The effort lasted one hour and forty-two minutes. During this time the APS had to keep the S-IVB stage stable. The system functioned without further problems during the rest of the mission. Like Apollo 13, at the end of the Saturn V mission the APS was fired twice, once between 006:04:02 MET and 006:05:22 MET, to put space between the stage and the CSM-LM combination, and once between 008:59:59 MET and 009:04:11 MET for the actual flight maneuver to cause the S-IVB to crash onto the moon near the Apollo 12 landing site.

After it was unloaded from the barge, the S-II stage was taken to the Vehicle Assembly Building (in background, together with the Launch Control Center), where the stacking of stages took place.

SA 510 (APOLLO 15)

S A 510 was the twenty-fifth booster of the Saturn family and the tenth Saturn V. It was the ninth manned Saturn of the

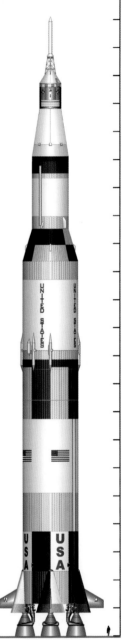

Saturn V (SA-510)
(NL : Apollo 15)

Mission Data and Flight Information SA 510	
Official Mission Designation	Apollo 15
Saturn Mission Designation	Apollo Saturn (SA) 510
Payload	Apollo CSM 112 and LM 10
Launch Site	LC-39A
Payload Weight	103,595 pounds
Arrival of S-IC at the Cape	July 6, 1970
Arrival of S-II at the Cape	May 18, 1970
Arrival of S-IVB at the Cape	June 13, 1969
Instrument Unit at the Cape	June 26, 1970
Stacking S-IC	July 8, 1970
Stacking S-II	September 15, 1970
Stacking S-IVB	September 16, 1970
Stacking Instrument Unit	September 17, 1970
Stacking Apollo and LM	May 8, 1971
Rollout to Launch Pad 39A	May 11, 1971
Start of Final Countdown	July 24, 1971, 23:00
S-IC Ignition Command	July 26, 1971, 13:33:51
Range Zero/Launch	13:34:00/000:00:00 MET
S-IC Cutoff Center Engine	000:02:16 MET
S-IC Cutoff Outer Engines	000:02:40 MET
S-II Engine Ignition	000:02:43 MET
Jettison S-II Interstage	000:03:11 MET
Escape Tower Separation	000:03:16 MET
S-II Outboard Engine Cutoff	000:09:09 MET
S-IVB First Burn Phase	000:09:13 MET
S-IC Impact in Water	000:09:21 MET
Position S-IC	29°42'N, 73°39'W
S-II Outer Engine Cutoff	000:09:09 MET
S-IVB Ignition Start	000:09:20 MET
End of First Burn S-IVB	000:11:35 MET
S-II Impact in Water	000:19:44 MET
Start of Second Burn Phase S-IVB	002:50:03 MET
End of Second Burn Phase S-IVB	002:55:54 MET
CSM Separation from S-IVB	003:22:27 MET
CSM Docking with LM/S-IVB	003:33:50 MET
CSM/LM Separation from S-IVB	004:18:01 MET
Ignition APS for Lunar Impact	010:00:01 MET
APS End of Burn	010:01:12 MET
S-IVB Impact on the Moon	079:24:43 MET
Lunar Coordinates S-IVB 509	1°57'S, 11°81'W

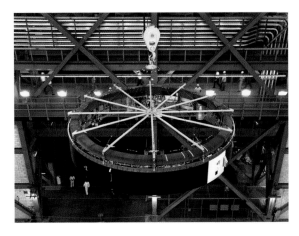

Stacking of the Instrument Unit with the third stage of SA 510 took place on September 17, 1970.

Apollo program and the eighth manned flight of a Saturn V.

SA 510 put astronauts David Scott, Alfred Worden, and James Irwin on the way to the fourth landing on the moon.

The assembly of SA 510 for the Apollo 15 mission began on Wednesday, July 8, 1970, in High Bay 3 of the Vehicle Assembly Building. Apollo spacecraft CSM 112 and Lunar Module 10 were mounted on the tip of the rocket on May 8, 1971. On December 15, 1969, Mobile Launcher 3 moved the rocket to Launch Pad 39A. Firing Room Number 1 in the launch control building at the Cape was used for this launch.

An irregularity occurred during the separation of the S-IC stage from the S-II stage, when the F-1 engine of the S-IC did not immediately shut down and accelerated for about four seconds with a certain residual thrust. This created the danger of a collision with the S-II stage, which fortunately did not happen. All subsequent separation procedures and burn maneuvers proceeded normally. The ullage case was separated at 000:29:21 MET.

As on all moon missions from Apollo 13, at the end of the STC mission the APS was ignited twice. The first maneuver, between 005:46:01 MET and 005:52:02 MET, was also to create space between the stage and the CSM-LM combination. The fine tuning of the collision course with the moon followed about four hours later. S-IVB 510 impacted the moon near the Apollo 14 landing site.

Significant figures in the creation of the Apollo program and senior NASA officials watch the launch of Apollo 15 in the launch control room of Kennedy Space Center. Far left, the longtime program head of the Apollo program, Gen. Sam Phillips; second from left, Wernher von Braun, longtime leader of the Marshall Space Flight Center, and in that post responsible for the development of the Saturn rockets; NASA administrator James Fletcher, who had just been appointed to that position at the time of the Apollo launch; and far right NASA's deputy administrator George Lowe.

SA 511 (APOLLO 16)

S A 511 was the twenty-sixth booster of the Saturn family and the eleventh Saturn V. It was the tenth manned Saturn of the Apollo program and the ninth manned flight of a Saturn V.

Mission Data and Flight Information SA 511	
Official Mission Designation	Apollo 16
Saturn Mission Designation	Apollo Saturn (SA) 511
Payload	Apollo CSM 113 and LM 11
Launch Site	LC-39A
Payload Weight	107,233 pounds
Arrival of S-IC at the Cape	September 17, 1971
Arrival of S-II at the Cape	September 30, 1970
Arrival of S-IVB at the Cape	July 1, 1970
Instrument Unit at the Cape	September 29, 1970
Stacking S-IC	September 21, 1971
Stacking S-II	October 1, 1971
Stacking S-IVB	October 5, 1971
Stacking Instrument Unit	October 6, 1971
Stacking Apollo and LM	December 8, 1971
Rollout to Launch Pad 39A	February 9, 1972
Start of Final Countdown	April 15, 1972, 03:54
S-IC Ignition Command	April 16, 1972, 17:53:51
Range Zero/Launch	17:54:00/000:00:00 MET
S-IC Cutoff Center Engine	000:02:18 MET
S-IC Cutoff Outer Engines	000:02:42 MET
S-II Engine Ignition	000:02:44 MET
Jettison S-II Interstage	000:03:14 MET
Escape Tower Separation	000:03:20 MET
S-IC Impact in Water	000:09:07 MET
Position S-IC	30°12'N, 74°09'W
S-II Inner Engine Cutoff	000:07:42 MET
S-II Outer Engine Cutoff	000:09:20 MET
S-IVB Ignition Start	000:09:21 MET
Start of First Burn Phase S-IVB	000:09:24 MET
End of First Burn Phase S-IVB	000:11:46 MET
S-II Impact in Water	000:20:02 MET
Start of Second Burn Phase S-IVB	002:33:37 MET
End of Second Burn Phase S-IVB	002:39:18 MET
CSM Separation from S-IVB	003:04:59 MET
CSM Docking with LM/S-IVB	003:21:53 MET
CSM/LM Separation from S-IVB	003:59:15 MET
Ignition APS for Lunar Impact	005:40:07 MET
APS End of Burn	005:41:01 MET
S-IVB Impact on the Moon	075:08:04 MET
Lunar Coordinates S-IVB 509	1°03'S, 23°08'W

Launch of Apollo 16 carried by SA 511, on April 16, 1972.

The S-IVB stage for Apollo 16 is unloaded from the Aero Spacelines Super Guppy at Cape Canaveral on July 1, 1970.

SA 511 put astronauts John Young, Ken Mattingly, and Charles Duke on their way to the fifth landing on the moon.

The assembly of SA 511 for the Apollo 15 mission began on Wednesday, September 21, 1971, in High Bay 3 of the Vehicle Assembly Building. Apollo spacecraft CSM 112 and Lunar Module 10 were mounted on the tip of the rocket on December 8, 1971. On December 13, 1971, Mobile Launcher 3 moved the rocket to Launch Pad 39A. As the result of a leak in the Apollo Service Module's attitude control system, on January 27, 1972, SA 511 was returned to the VAB

for repairs. The second roll-out took place on February 9, 1972. Firing Room 1 in the Cape's launch control center monitored the launch.

The ullage case was jettisoned at 000:09:32 MET.

As on all moon missions from Apollo 13, at the end of the Saturn V mission the APS was ignited twice. The first maneuver, between 004:18:08 MET and 004:19:29 MET, was also to create space between the stage and the CSM-LM combination. The fine tuning of the collision course with the moon followed about eighty minutes later.

The F-1 engines underwent at least three burn tests before they were qualified for a flight mission. First were two individual tests of 40 and 165 seconds duration on the manufacturer's test stand. Then they were installed in the stage and, together with the other four engines, were tested on the B-2 test stand in Mississippi for 125 seconds. In the end the total running time was thus on average 495 seconds, for the operating time during the flight mission was about 165 seconds. The testing times differed considerably, however. Including various burn tests, the engine with the longest individual running time, including the subsequent mission, logged a total running time of 800 seconds.

The Apollo 16 spacecraft Casper with the adapter enclosing the Lunar Module Orion is attached to AS 511 on December 8, 1971.

131

SA 512 (APOLLO 17)

S A 512 was the twenty-seventh booster of the Saturn family and the twelfth Saturn V. It was the eleventh and last manned Saturn of the Apollo program and the tenth and last manned flight of a Saturn V. It was the second night launch of the Apollo program

The launch of Saturn SA 512 with Apollo 17 was the second night launch of the Apollo program, and the first manned night launch of the entire US manned spaceflight program.

Mission Data and Flight Information SA 512	
Official Mission Designation	Apollo 17
Saturn Mission Designation	Apollo Saturn (SA) 512
Payload	Apollo CSM 114 and LM 12
Launch Site	LC-39A
Payload Weight	107,144 pounds
Arrival of S-IC at the Cape	May 11, 1972
Arrival of S-II at the Cape	October 27, 1970
Arrival of S-IVB at the Cape	December 21, 1970
Instrument Unit at the Cape	June 7, 1972
Stacking S-IC	May 15, 1972
Stacking S-II	May 19, 1972
Stacking S-IVB	June 20, 1972
Stacking Instrument Unit	June 23, 1972
Stacking Apollo and LM	August 23, 1972
Rollout to Launch Pad 39A	August 28, 1972
Start of Final Countdown	December 5, 1972, 12:53
S-IC Ignition Command	December 7, 1972, 05:32:51
Range Zero/Launch	05:33:00/000:00:00 MET
S-IC Cutoff Center Engine	000:02:19 MET
S-IC Cutoff Outer Engines	000:02:41 MET
S-II Engine Ignition	000:02:45 MET
Jettison S-II Interstage	000:03:13 MET
Escape Tower Separation	000:03:19 MET
S-IC Impact in Water	000:09:12 MET
Position S-IC	28°13′N, 73°53′W
S-II Inner Engine Cutoff	000:07:41 MET
S-II Outer Engine Cutoff	000:09:20 MET
S-IVB Ignition Start	000:09:21 MET
Start of First Burn Phase S-IVB	000:09:24 MET
Ullage Case Separation	000:09:32 MET
End of First Burn Phase S-IVB	000:11:43 MET
S-II Impact in Water	000:19:57 MET
Start of Second Burn Phase S-IVB	003:12:37 MET
End of Second Burn Phase S-IVB	002:18:28 MET
CSM Separation from S-IVB	003:42:28 MET
CSM Docking with LM/S-IVB	003:57:11 MET
CSM/LM Separation from S-IVB	004:45:02 MET
Ignition APS for Lunar Impact	011:15:00 MET
APS End of Burn	011:16:42 MET
S-IVB Impact on the Moon	086:49:52 MET
Lunar Coordinates S-IVB 512	4°21′S, 12°31′W

Saturn SA 512 with Apollo 17 during transport to the launch pad at Kennedy Space Center on August 28, 1972.

and the first manned night launch ever in the American manned space program. SA 512 put astronauts Eugene Cernan, Ronald Evans, and Harrison Schmitt on their way to the sixth and last landing on the moon of the Apollo program.

The assembly of SA 512 for the Apollo 17

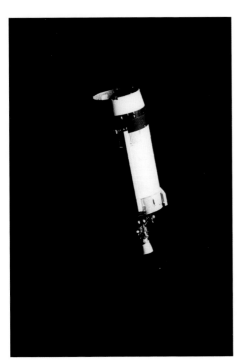

mission began on Monday, May 15, 1972, in High Bay 3 of the Vehicle Assembly Building. Apollo spacecraft CSM 114 and Lunar Module 12 were mounted on the tip of the rocket on August 23, 1972. On August 28, 1972, Mobile Launcher 2 moved the rocket to Launch Pad 39A. Firing Room 1 in the Cape's launch control center monitored the launch.

Apollo 17 launched at 1233 Eastern Time. The launch was delayed by two hours and forty minutes due to an automatic cutoff in the launch sequencer at the T-30 second mark in the countdown. The error was quickly identified as a minor technical problem and repaired. The launch clock was set back to T-22 minutes and the countdown was resumed.

As on all moon missions from Apollo 13, at the end of the Saturn V mission the APS was ignited twice. The first maneuver, between 006:10:00 MET and 006:11:38 MET, was also to create space between the stage and the CSM-LM combination. The fine tuning of the collision course with the moon followed about four hours later between 11:15:00 MET and 11:16:42 MET.

S-IVB-512 was the last S-IVB stage to fly to the moon. This photo was taken about five hours after launch from Kennedy Space Center, and was taken by the Apollo 17 astronauts.

SA 513 (SKYLAB 1)

S A 513 was the twenty-eighth booster of the Saturn family and the thirteenth and last Saturn V ever used. While units SA 514 to SA 516 were completed, they were never used operationally and today are museum pieces. Skylab 1 was the first mission of the Skylab program and the third and last unmanned flight of a Saturn V. SA 513 delivered the space station Skylab 1 into earth orbit.

The assembly of SA 513 for the Skylab 1 mission began on Thursday, August 2, 1972 in High Bay 2 of the Vehicle Assembly Building. Installation of the space laboratory on the rocket took a long period of time. It began at the end of September 1972, and went on until January 30, 1973. On April 26 Mobile Launcher 2 drove the rocket to Launch Complex 39A. Firing Room 1 in the Cape's launch control center monitored the launch.

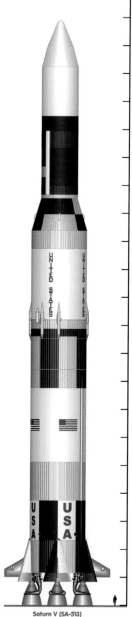

Saturn V (SA-513)
(NL : Skylab 1)

Mission Data and Flight Information SA 513	
Official Mission Designation	Skylab 1
Saturn Mission Designation	Apollo Saturn (SA) 513
Payload	Skylab
Launch Site	LC-39A
Payload Weight	170,000 pounds
Arrival of S-IC at the Cape	July 26, 1972
Arrival of S-II at the Cape	January 1, 1971
Arrival of S-IVB at the Cape	September 22, 1972
Instrument Unit at the Cape	October 27, 1972
Stacking S-IC	August 2, 1972
Stacking S-II	September 20, 1972
Stacking Instrument Unit	January 30, 1973
Stacking Skylab	January 30, 1973
Rollout to Launch Pad 39A	April 26, 1973
S-IC Ignition Command	May 14, 1973, 17:29:51
Range Zero/Launch	17:30:00/000:00:00 MET
Skylab Damaged	000:01:03 MET
S-IC Cutoff Center Engine	000:01:03 MET
S-IC Cutoff Outer Engines	000:02:19 MET
S-II Engine Ignition	000:02:45 MET
S-II Entry into Earth's Atmosphere	January 11, 1975
Skylab Separation from S-II	003:42:28 MET
S-II Cutoff Außentriebwerke	000:09:48 MET
S-II Eintritt in Erdatmosphäre	11. Januar 1975
Skylab-Trennung von S-II	003:42:28 MET

A variant of the Saturn V that is almost forgotten today. This two-stage version flew just once, carrying America's first and only national space station—Skylab.

Sixty-three seconds after the Saturn left the launch pad, in the zone of maximum dynamic loads on the rocket, one of the space station's two solar generators together with the meteorite shield and the thermal cover were torn off. The Saturn nevertheless placed the badly-damaged space station into orbit. In a dramatic rescue operation, Skylab's first crew succeeded in activating the space station despite the severe damage.

To give the station some protection from the rough stage separation, customary for the Saturn V, the burn termination sequence of the S-IC stage was modified for this last mission. Instead of shutting down the center engine first and then the four remaining rocket engines, on this mission

Skylab 1 lifts off on May 14, 1973.

the center engine was first shut down, then two of the remaining engines followed by the final two.

The S-II second stage of the SA 513 was the only one to reach earth orbit in the entire Apollo program. It did not reenter earth's atmosphere until January 11, 1975. Together with the burnt-out second stage, the total weight of the payload delivered to orbit was 113.4 tons.

The outer hull of the Skylab space laboratory was essentially S-IVB stage 212, whose tank section had been converted into the space laboratory. Before it was put into space it was loaded with supplies and equipment for a total of three manned missions.

Conversion work on Launch Complex 39A for use in the future Shuttle program began shortly before the launch of Skylab 1. The manned Skylab and ASTP launches therefore all took place from Launch Complex 39B.

SA 206 (SKYLAB 2)

Saturn SA 206 was the twenty-ninth booster of the Saturn family and the sixth Saturn IB. It was the second manned flight of a Saturn IB. SA 206 delivered astronauts Charles Conrad, Joseph Kerwin, and Paul Weitz to earth orbit for the first manned mission aboard the American space station Skylab.

SA 206 was the first Saturn IB to be assembled in the Vehicle Assembly Building. For the missions of Skylab 2, 3, and 4, and the ASTP mission, Mobile Launcher 1 had been fitted with a platform, the so-called Milkstool, which enabled it to use Launch Complex 39B at the Kennedy Space Center. SA 206 was thus the first Saturn IB to be launched from the Kennedy Space Center, and not from Cape Canaveral Air Base, from Launch Complex 34 or 37B, and the second Saturn ever to depart Launch Complex 39B for space.

SA 206 was also the first to use an improved, more powerful version of the H-1 first stage engine.

Launch of Skylab 2 at 0800 Eastern Time, on May 25, 1973.

Mission Data and Flight Information SA 207	
Official Mission Designation	Skylab 3
Saturn Mission Designation	Apollo Saturn (SA) 207
Payload	Apollo CSM 117
Launch Site	LC-39B
Payload Weight	44,368 pounds
Arrival of S-IB at the Cape	March 30, 1975
Arrival of S-IVB at the Cape	August 26, 1971
Instrument Unit at the Cape	May 8, 1973
Stacking S-IB	April 4, 1973
Stacking S-IVB	May 28, 1973
Stacking Instrument Unit	May 29, 1973
Stacking Apollo CSM	June 3, 1973
Rollout to Launch Pad 39B	June 11, 1973
S-IB Ignition Command	July 28, 1973, 11:16:39
Range Zero/Launch	11:16:42/000:00:00 MET
S-IB Cutoff Center Engine	000:02:18 MET
S-IB Cutoff Outer Engines	000:02:21 MET
S-IB/S-IVB Separation Command	000:02:22 MET
S-IVB Ignition Command	000:02:23 MET
Jettison S-IVB Ullage Case	000:02:34 MET
Separation Escape Tower	000:02:45 MET
S-IVB End of Burn	000:09:52 MET
Separation of CSM from S-IVB	000:16:52 MET
End of Mission S-IVB	007:30:00 MET

A Saturn IB first stage is assembled in the Vehicle Assembly Building at Kennedy Space Center for the first time. The photo was taken in late August 1972.

Stacking of SA 206 for the Skylab 2 mission began on Thursday, August 31, 1972, in High Bay 1 of the Vehicle Assembly Building. On February 26, Mobile Launcher 1 delivered Skylab to Launch Complex 39B.

SA 206 had a long history behind it prior to the launch. In 1967, the rocket had been assembled at Launch Complex 37B for the first (unmanned) flight of a Lunar Module. After the Apollo 1 fire, when the entire Apollo program was planned anew, the rocket had been disassembled, sent back to Michoud, and stored there.

The Skylab missions were the penultimate chapter in the history of the Saturn rockets. There were five Skylab missions, using one Saturn V and four Saturn IB rockets. Three of these Saturn IBs were used for manned shuttle missions to the space station.

Skylab's orbit was at an inclination of fifty degrees at an average altitude of 267 miles. It was thus at a similar inclination as the present-day International Space Station and also at a similar orbital altitude.

Skylab 2 was supposed to have launched on May 15, 1973, but after Skylab 1 was badly damaged during launch, an emergency plan to save the space laboratory had to be worked out first.

This photo illustrates well how it was possible to launch the much smaller Saturn IB from Launch Complex 39B, which had been designed for the Saturn V. Here the rocket is sitting on a pedestal more than ninety-eight feet high.

SA 207 (SKYLAB 3)

Saturn SA 207 was the thirtieth booster of the Saturn family and the seventh Saturn IB. It was the third manned flight by a Saturn IB. SA 207 delivered astronauts Alan Bean, Jack Lousma, and Owen Garriott into earth orbit for the second manned mission on board the American Skylab space station.

SA 207 was the second Saturn IB assembled in the Vehicle Assembly Building. Instrument Unit S-IU-208 was used on this mission. IU-207 flew on SA 208.

Stacking of SA 207 for the Skylab 3 mission began in High Bay 1 of the Vehicle Assembly Building on April 4, 1973. On June 11, Mobile Launcher 1 delivered Skylab 3 to Launch Complex 39B. Firing Room 3 at the Kennedy Space Center was chosen to control the launch.

One noticeable difference between the Skylab Saturn IB and those of the Apollo program was the painting of the first stage. Saturn IBs of the Apollo program had white-painted stages with

Launch of SA 207 for Skylab 3 mission on July 28, 1973.

Mission Data and Flight Information SA 207	
Official Mission Designation	Skylab 3
Saturn Mission Designation	Apollo Saturn (SA) 207
Payload	Apollo CSM 117
Launch Site	LC-39B
Payload Weight	44,368 pounds
Arrival of S-IB at the Cape	March 30, 1973
Arrival of S-IVB at the Cape	August 26, 1971
Instrument Unit at the Cape	May 8, 1973
Stacking S-IB	April 4, 1973
Stacking S-IVB	May 28, 1973
Stacking Instrument Unit	May 29, 1973
Stacking Apollo CSM	June 3, 1973
Rollout to Launch Pad 39B	June 11, 1973
S-IB Ignition Command	July 28, 1973, 11:16:39
Range Zero/Launch	11:16:42/000:00:00 MET
S-IB Cutoff Center Engine	000:02:18 MET
S-IB Cutoff Outer Engines	000:02:21 MET
S-IB/S-IVB Separation Command	000:02:22 MET
S-IIVB Engine Ignition Command	000:02:23 MET
S-IVB Ullage Case Jettison	000:02:34 MET
Separation Escape Tower	000:02:45 MET
S-IVB End of Burn	000:09:52 MET
Separation of CSM from S-IVB	000:16:52 MET
End of Mission S-IVB	007:30:00 MET

vertical black stripes. This scheme was abandoned for the first stages of the Skylab Saturn IBs. As these rockets were exposed to the sun on the Milkstool and therefore experienced greater heating than the earlier Saturn IBs launched from Launch Complexes 34 and 37B, the S-IVB stages were painted completely white to prevent heating of the liquid oxygen. On missions SA 201 and SA 202, the interstage area between the S-IVB stage and the spacecraft adapter was initially painted white. All subsequent Saturn IBs, however, had a black horizontal bar in this area.

The S IVB 207 shortly after achieving orbit, photographed by the crew of Skylab 3.

The Skylab space station. The central element on the left is nothing more than the S-IVB 212, which was modified for this purpose. The combined oxygen/hydrogen tank was modified to provide living quarters for the crew.

139

SA 208 (SKYLAB 4)

Saturn SA 208 was the thirty-first booster of the Saturn family and the eighth Saturn IB. It was the fourth manned flight by a Saturn IB. SA 208 delivered astronauts Gerald Carr, William Pogue, and Edward Gibson into earth orbit for the third manned mission on board the American Skylab space station. SA 208 was the third Saturn IB stacked in the Vehicle Assembly Building.

The stacking of SA 208 for the Skylab 4 mission began on July 31, 1973, in High Bay 1 of the Vehicle Assembly Building. On August 14, Mobile Launcher 1 transported Skylab 4 to Launch Complex 39B. Mission control for the launch was carried out from Firing Room 3 at the John F. Kennedy Space Center. Instrument Unit S-IU-207 was used on this flight.

Because of a failure in the attitude control system—one of Apollo Command and Service Module 117's four attitude control modules failed and a second developed an oxidizer leak—during

The launch of Saturn IB SA 208 on the Skylab 4 mission, on November 16, 1973.

the mission of Skylab 3 preparations were begun for a rescue mission for the crew of Skylab 3. It was planned that astronauts Vance Brand and Don Lind would pick up the crew in the event of an emergency. Saturn IB SA 209 was prepared for this possible mission. The spacecraft and rocket were combined, taken to Launch Complex 39B and held in readiness there, but then it was decided that Apollo Command and Service Module 117 itself could manage the return flight with two inactive attitude control modules. Just two years later Saturn IB SA 209 was again held ready as the back-up vehicle, this time for the ASTP mission. In the end SA 209 never flew and for decades has been on display at the Kennedy Space Center.

Mission Data and Flight Information SA 208	
Official Mission Designation	Skylab 4
Saturn Mission Designation	Apollo Saturn (SA) 208
Payload	Apollo CSM 118
Launch Site	LC-39B
Payload Weight	45,966 pounds
Arrival of S-IB at the Cape	June 20, 1973
Arrival of S-IVB at the Cape	November 4, 1971
Instrument Unit at the Cape	June 12, 1973
Stacking S-IB	July 31, 1973
Stacking S-IVB	August 1, 1973
Stacking Instrument Unit	May 29, 1973
Stacking Apollo CSM	August 6, 1973
Rollout to Launch Pad 39B	August 14, 1973
S-IB Ignition Command	November 16, 1973, 14:01:20
Range Zero/Launch	14:01:23/000:00:00 MET
S-IB Cutoff Center Engine	000:02:18 MET
S-IB Cutoff Outer Engines	000:02:21 MET
S-IB/S-IVB Separation Command	000:02:22 MET
S-IIVB Engine Ignition Command	000:02:23 MET
S-IVB Ullage Case Jettison	000:02:34 MET
Separation Escape Tower	000:02:45 MET
S-IVB End of Burn	000:09:52 MET
Separation of CSM from S-IVB	000:16:52 MET
End of Mission S-IVB	007:30:00 MET

Production of SA 208's
S-IB stage at Michoud.

This photo also clearly shows
that the Skylab space laboratory
was nothing but a modification
of an S-IVB stage with some
added equipment. The S-IVB
stage 212 was the last unit ever
built.

SA 210 (APOLLO-SOYUZ TEST PROJECT)

S aturn SA 210 was the thirty-second and last booster of the Saturn family to be used operationally and the ninth Saturn IB to fly. SA 209 had also been prepared for a launch, for potential rescue flights for the Skylab 3 and 4 and ASTP missions, but was not used. SA 210 was the fifth and last manned flight of a Saturn IB. The booster delivered astronauts Thomas Stafford, Deke Slayton, and Vance Brand into orbit for a rendezvous with Soyuz 19. SA 210 was the fifth and last Saturn IB stacked in the Vehicle Assembly Building.

The stacking of SA 210 for the ASTP mission began on Monday, January 13, 1975, in High Bay 1 of the Vehicle Assembly Building. Mobile Launcher 1 transported the ASTP Apollo to Launch Complex 39B on March 24. Firing Room 3 at the Kennedy Space Center was

Mission Data and Flight Information SA 210	
Official Mission Designation	ASTP
Saturn Mission Designation	Apollo Saturn (SA) 210
Payload	Apollo CSM 111
Launch Site	LC-39B
Payload Weight	36,994 pounds
Arrival of S-IB at the Cape	April 22, 1974
Arrival of S-IVB at the Cape	November 6, 1972
Instrument Unit at the Cape	May 14, 1974
Stacking S-IB	January 13, 1975
Stacking S-IVB	January 14, 1975
Stacking Instrument Unit	January 16, 1975
Stacking Apollo CSM	March 19, 1975
Rollout to Launch Pad 39B	March 24, 1975
S-IB Ignition Command	July 15, 19:49:57
Range Zero/Launch	19:50:00/000:00:00 MET
S-IB Cutoff Center Engine	000:02:18 MET
S-IB Cutoff Outer Engines	000:02:21 MET
S-IB/S-IVB Separation Command	000:02:22 MET
S-IIVB Engine Ignition Command	000:02:23 MET
S-IVB Ullage Case Jettison	000:02:34 MET
Separation Escape Tower	000:02:45 MET
S-IVB End of Burn	000:09:52 MET
Separation of CSM from S-IVB	000:16:52 MET
End of Mission S-IVB	007:30:00 MET
„End-of-Mission S-IVB"	007:30:00 MET

Saturn IB
(SA 210)

A view of the interior of the Vehicle Assembly Building, where after stacking the Saturn IB the fins are being mounted on SA 210's first stage. The photo was taken on March 11, 1975.

used for launch mission control.

The abbreviation ASTP stood for Apollo–Soyuz Test Project. It was the first joint mission with the former Soviet Union. ASTP was the last American manned mission until April 12, 1981, when the Space Shuttle *Columbia* took off on mission STS-1.

As on the Apollo moon flights, after launch the pilot of the Command Module, in this case astronaut Vance Brand, had to turn the spacecraft and dock with the docking adapter, which was housed in the payload segment above the S-IVB stage, and extract it. This was the only Saturn IB mission in which a secondary payload was carried.

Mobile Launcher 1 took SA 210 to Launch Complex 39B, on March 24, 1975.

Project Apollo: The Early Years, 1960–1967. Eugen Reichl. In May 1961, American President John F. Kennedy committed the nation to carrying out a manned landing on the moon before the end of the decade. This volume covers the early years of the Apollo program, still the most significant space effort in the history of mankind.
Size: 6"x9" • 127 color images • 144pp.
ISBN: 978-0-7643-5174-7 • hard cover • $19.99

Project Apollo: The Moon Landings, 1968–1972. Eugen Reichl. This book brings the later years of the Apollo era to new life with details including the test flights in Earth's orbit; the first orbits of the moon; the legendary Apollo 11 mission; the drama of Apollo 13; and Apollo 17, the last manned moon flight in 1972.
Size: 6"x9" • 130 color and b/w images • 144pp.
ISBN: 978-0-7643-5375-8 • hard cover • $19.99

Project Gemini. Eugen Reichl. This second volume in the "America in Space" series continues the history of the American manned space program. Beginning in 1964, two unmanned and ten manned flights took place in the Gemini program. All Project Gemini missions are discussed, including details on all craft and the astronauts involved. Superb color, archival images, cutaways, and plans are also included.
Size: 6" x 9" • 132 color images • 144pp.
ISBN: 978-0-7643-5070-2 • hard • $19.99

Project Mercury. Eugen Reichl. This book is a concise, detailed history of America's first steps into space. Project Mercury was America's entry into the manned spaceflight program. All missions in Project Mercury are discussed, including details on all craft and the astronauts involved. Superb color, archival images, cutaways, and plans are also included.
Size: 6" x 9" • 89 color images • 144pp.
ISBN: 978-0-7643-5069-6 • hard • $19.99

 SCHIFFERPUBLISHING

 SCHIFFERPUBLISHING

 SCHIFFERBOOKS

 SCHIFFERBOOKS

 Schiffer Publishing, Ltd., 4880 Lower Valley Road, Atglen, PA 19310
Phone: (610) 593-1777; Fax: (610) 593-2002; E-mail: info@schifferbooks.com
Ordering Hours: 8:30 a.m. – 5:30 p.m. Eastern Time, Monday–Friday

Find digital catalogs on our website, www.schifferbooks.com and Edelweis
If you would like to receive other catalogs,
please contact our customer service team
at customercare@schifferbooks.com or (610) 593-1777.